Springer Series in *Materials Science* 30

Edited by H. Sakaki

Springer

Berlin
Heidelberg
New York
Barcelona
Budapest
Hong Kong
London
Milan
Paris
Santa Clara
Singapore
Tokyo

Springer Series in *Materials Science*

Advisors: M.S. Dresselhaus · H. Kamimura · K.A. Müller
Editors: U. Gonser · R.M. Osgood, Jr. · M.B. Panish · H. Sakaki
Managing Editor: H.K.V. Lotsch

Volumes 1–28 are listed at the end of the book.

Kenichi Iga   Susumu Kinoshita

Process Technology for Semiconductor Lasers

Crystal Growth and Microprocesses

With 115 Figures and 20 Tables

 Springer

Professor Dr. Kenichi Iga
Dr. Susumu Kinoshita

Tokyo Institute of Technology
Precision and Intelligence Laboratory
4259 Nagatsuta, Midori-ku
Yokohama 226, Japan

Series Editors:

Prof. Dr. U. Gonser

Fachbereich 12.1, Gebäude 22/6
Werkstoffwissenschaften
Universität des Saarlandes
D-66041 Saarbrücken, Germany

M. B. Panish, Ph. D.

AT&T Bell Laboratories
600 Mountain Avenue
Murray Hill, NJ 07974-2070, USA

Prof. R. M. Osgood, Jr.

Microelectronics Science Laboratory
Department of Electrical Engineering
Columbia University
Seeley W. Mudd Building
New York, NY 10027, USA

Prof. H. Sakaki

Institute of Industrial Science
University of Tokyo
7-22-1 Roppongi, Minato-ku
Tokyo 106, Japan

Managing Editor:

Dr.-Ing. Helmut K. V. Lotsch

Springer-Verlag, Tiergartenstrasse 17
D-69121 Heidelberg, Germany

Library of Congress Cataloging-in-Publication Data

Iga, Ken'ichi, 1940-
    Process technology for semiconductor lasers : crystal growth and
microprocesses / Kenichi Iga, Susumu Kinoshita.
        p.   cm. -- (Springer series in materials science)
    Includes bibliographical references and index.

    1. Semiconductor lasers--Design and construction. 2. Laser
materials. 3. Crystal growth. 4. Epitaxy.   I. Kinoshita, Susumu,
1959-   . II. Title. III. Series.
TA1700.I49 1996
621.36'6--dc20   95-48999
                                                    CIP

ISBN-13: 978-3-642-79578-7   e-ISBN-13: 978-3-642-79576-3
DOI: 10.1007/978-3-642-79576-3

© Springer-Verlag Berlin Heidelberg 1996

Softcover reprint of the hardcover 1st edition 1996

Typesetting: PS™ Technical Word Processor
SPIN: 10009470   54/3144-5 4 3 2 1 0 – Printed on acid-free paper

Preface

This book presents fundamental process technologies for semiconductor lasers. Process technologies regarding semiconductor-laser fabrication are discussed with emphasis mainly on crystal-growth methods such as liquid phase epitaxy, metalorganic chemical vapor deposition, molecular beam epitaxy, chemical beam epitaxy and so on. Representative semiconductor-laser performances are described, and the latest topics including mode control and surface-emitting lasers are introduced.

The purpose of this book is to describe the design principles of semiconductor lasers, mainly from the fabrication point of view. We describe advanced technologies of semiconductor lasers, particularly device design and fabrication, and semiconductor-laser characteristics.

In Chap. 1 we start out with the history of semiconductor-laser development and applications. Then in Chap. 2 materials for use in semiconductor lasers ranging from short to long wavelengths are reviewed. Chapter 3 touches on the basic design principles of semiconductor-laser devices. In Chap. 4 we review epitaxies for laser fabrication. Chapter 5 is devoted to the technology for liquid phase epitaxy and Chap. 6 is on vapor phase and beam epitaxies. Chapter 7 introduces characterizations of laser materials. In Chap. 8 the fabrication and characteristics of semiconductor lasers are detailed. Mode-control techniques are presented in Chap. 9, and finally surface-emitting lasers are introduced in Chap. 10.

This book will be beneficial for scientists and manufacturers in institutes and industries. In addition, professors and students in graduate courses at universities can utilize this book as a textbook or reference to research and advanced studies.

The authors would like to acknowledge Professor Emeritus Yasuhara Suematsu for encouragement and Professor Fumio Koyama for discussion. Most of the art works redrawn for the present book are due to the staff and students at the authors' laboratory. We also thank Ms. Fumiyo Matsunaga, secretary of our labotary, and Tohru Honda for helping in the arrangement of the manuscript. Lastly, we thank Helmut K.V. Lotsch of Springer-Verlag for cooperation in the preparation of this book.

Yokohama   *K. Iga*
October 1995   *S. Kinoshita*

Contents

1. Introduction

In this chapter we will outline the theory of semiconductor laser devices, and summarize the associated fabrication technology and their importance in opto-electronics. A brief history of semiconductor lasers will be given along with a description of current applications.

1.1 Outline of Semiconductor Laser Theory

It is possible for a semiconductor to emit light with a wavelength inversely proportional to the band-gap energy E_g. When E_g is expressed in units of eV, the wavelength λ_0 is given as

$$\lambda_0 = 1.2398/E_g \quad [\mu m], \tag{1.1}$$

since the emission wavelength depends upon the band-gap energy, the emitted light can vary through wide ranges of wavelength by changing the composition of the compound semiconductor.

A laser oscillator is just like an electronic oscillator in that it requires a resonator in which feedback of light can occur in the same space. A Fabry-Perot interferometer consisting of two reflecting, plane mirrors, as shown in Fig.1.1, is a well-known optical resonator. The light is reflected and confined between the two reflecting mirrors to produce a standing wave. Stimulated emission occurs with the same phase as that of the standing wave.

If the amplification of light is equal to the total loss, i.e., the sum of the absorption and reflection losses, and if a resonance wavelength exists within the gain bandwidth, then oscillation will take place. The necessary gain to reach the threshold of oscillation g_{th} is written as

$$g_{th} = \alpha + \frac{1}{2L} \ln\left(\frac{1}{R_1 R_2}\right). \tag{1.2}$$

Here, we have assumed that the optical gain and loss are g [cm^{-1}] and α [cm^{-1}], respectively, the reflectivity of the reflecting mirrors are R_1 and R_2, and the cavity length is L [cm].

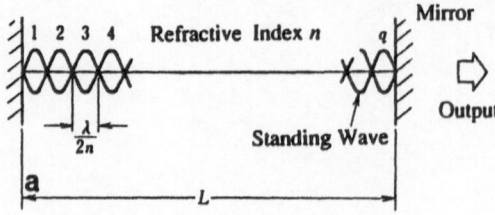

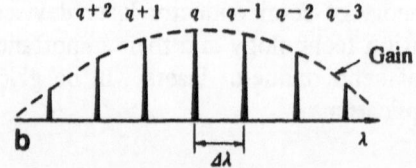

Fig.1.1a,b. Standing waves and spectra in a laser cavity

For example, light absorption, mostly due to *free-electron absorption*, exists in a semiconductor medium, and α is nearly equal to 10 cm^{-1}. When the length of the resonator, L, is 0.03 cm and the reflectivity $R_1 = R_2 = 0.3$ by taking advantage of the Fresnel reflection between the semiconductor (refractive index: 3.5) and air, g_{th} must be approximately 50 cm^{-1}.

The spectral linewidth of a laser is remarkably narrower than that of luminescence or spontaneous emission above threshold. Let us assume that n is the refractive index of the medium in the resonator. Multiplying the resonance wavelength λ by the integer q to complete one round-trip of the optical path with length 2nL, then λ is given as follows [1.1]

$$\lambda = 2nL/q .\tag{1.4}$$

Because the magnitude of q is large, the wavelength changes only by $\Delta\lambda$ even if q changes by 1. Accordingly $\lambda \gg |\Delta\lambda|$, and the following equation is obtained

$$|\Delta\lambda| = \frac{\lambda^2}{2n_{eff}L} ,\tag{1.5}$$

where n_{eff} is an effective index of refraction, at $\lambda = \lambda_0$, which accounts for the wavelength dispersion, and is defined by

$$n_{eff} = n_0\left[1 - \frac{\lambda_0}{n_0}\frac{\partial n}{\partial\lambda}\bigg|_{\lambda=\lambda_0}\right] .\tag{1.6}$$

When the spectral gain width is large, oscillation is possible at various wavelengths. Modes with different wavelengths are called *longitudinal modes*. In the direction perpendicular to the propagation of light, the light may have various forms of beam shapes or *transverse modes*. For a wide range of applications a single transverse mode is desirable. The nominal wavelength of a laser is determined by the band-gap energy of the materials, and the actual lasing wavelength depends on the laser resonator as in (1.4). Within a gain spectrum the wavelength can be tuned by varying the length or resonance wavelength of the resonator.

1.2 Semiconductor Lasers in Opto-electronics

A semiconductor laser can be viewed both as a semiconductor device and as a laser. Like other semiconductor devices, the semiconductor laser can operate with a supply voltage of ≈ 2 V. Its intensity can normally be modulated up to several GHz simply by controlling the injection current. Laser chips measure approximately $200 \times 250 \times 100 \ \mu m^3$, which is $1/10$ to $1/100$ the size of gas lasers. Semiconductor lasers are more advantageous than Light Emitting Diodes (LEds) because they emit coherent light.

Coherent light is a remarkable characteristic of laser radiation. Two aspects of coherence are *temporal coherence* and *spatial coherence*. The former represents monochromaticity (or single frequency) and the extent of a wave packet. Although the laser light is said to be coherent, it is not completely so. Since the wave packet is finite or has fluctuations in amplitude and phase it usually has some amount of linewidth. This linewidth is not the gain-width, as described previously, but the narrow width around a resonance frequency which can be related to quantum noise.

When the spectral linewidth is $\delta\lambda_c$, the distance ℓ_c along which interference takes place in the light propagation, is written, referring to Fig. 1.2, as [1.2]

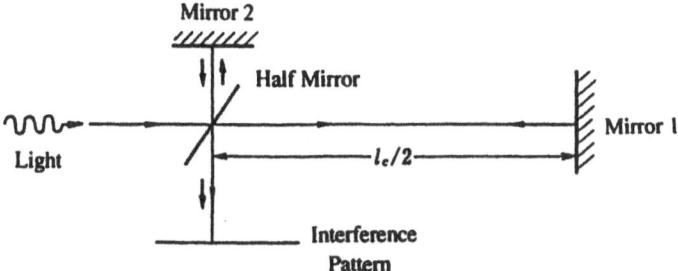

Fig.1.2. Measurement of temporal coherence by an interferometer

$$\ell_c = \frac{\lambda^2}{2\delta\lambda_c} . \tag{1.7}$$

The distance ℓ_c is longer than 10 to 100 cm in usual semiconductor lasers. Coherent optical communications utilizing such phase information represent a good application of the temporal coherence of laser beams.

On the other hand, spatial coherence means the degree of interference of a beam in its transverse direction. The interference can be seen intuitively by making interference patterns using two pinholes or slits. The maximum interference length in the transverse direction of light propagation is called the *coherence length* x_c. Laser beams are nearly coherent approximately 1 mm after emission from the laser and if broadened by an excellent-quality lens, etc., the coherence length can be made longer. However, the length cannot be made extremely large due to the aberration or inhomogeneity of the optics used. Let us consider focusing by a lens. Assuming that the coherence length in the transverse direction is x_c, the diameter ΔD of the beam spot focused by an aberration-free lens is given as [1.2]

$$\Delta D = 2.44\lambda f/x_c \tag{1.8}$$

where x_c must be smaller than the diameter D of the lens. If $f/D = 1$, $\lambda = 0.78$ μm and $x_c = D$, the diameter of the beam spot is $\Delta D = 1.9$ μm.

The diffraction angle $2\Delta\theta$ of a beam emitted into space can be related to the coherence length x_c, and written as [1.2]

$$2\Delta\theta = 1.22\lambda/x_c . \tag{1.9}$$

Laser radiation produces a highly directional beam of light since a large value of x_c is obtainable. Owing to their important and unique properties, semiconductor lasers are essential for optical communication and opto-electronic devices.

1.3 Necessary Technology for Semiconductor Lasers

The necessary technology for building a semiconductor laser is as follows:

- Formation of an optical resonator, and
- carrier injection or pumping.

Most of the necessary technology is closely related to crystal growth, since the optical resonator usually consists of a waveguide and two mirror facets or gratings, and carriers are injected through a p-n junction in a semicon-

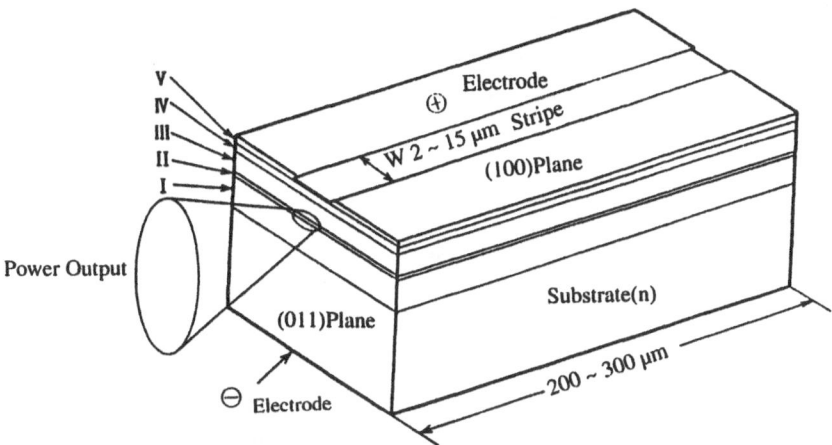

Fig.1.3. A double-heterostructure semiconductor laser

ductor. A double heterostructure contains a narrow band-gap, active layer sandwiched between large band-gap semiconductors. It represents an important semiconductor laser because it functions as a waveguide by confining carriers in the active layer or resonator.

Figure 1.3 is a schematic diagram of a semiconductor laser. The Double-Heterostructure (DH) laser [1.3] is effectively a waveguide because, when carriers are injected into a p-n junction, they are confined to the active layer by the surrounding heterostructure layers. The DH laser is formed on the substrate by epitaxial crystal growth, and the two mirror facets of the waveguide are obtained by cleaving the material. Lasing occurs when a minimum number of injected carriers are present, i.e., when the threshold current-density condition has been reached. Epitaxial crystal growth of DH lasers is one of the most important techniques in semiconductor-laser technology. This monograph is therefore dedicated to semiconductor lasers and the epitaxial crystal growth of the required materials.

1.4 Brief History of Semiconductor Lasers

The semiconductor laser was developed in 1962 and has become a crucial device for optical communications, optical-discs players and other optoelectronic systems. We can classify the development of semiconductor lasers into five stages, as suggested in Table 1.1.

The first stage ranges from the birth of a homojunction semiconductor injection laser; in 1962 four groups in the US (GE [1.4], MIT [1.5], IBM [1.6] and the University of Illinois [1.7]) succeeded in semiconductor-laser

Table 1.1. Development of semiconductor lasers

Dev. stage	Time period	Laser type	Main concepts
I	1962 ÷ 1969	GaAs homojunction	Direct modulation
II	1970 ÷ 1976	GaAlAs double-heterojunction	Transverse-mode control
III	1977 ÷ 1984	1 μm range GaInAsP group	DSM operation[a]
IV	1984 ÷ 1990	Short wavelength	Low noise
V	1990 →	Wide spectral range	Quantum well Coherence control Surface emission Red, green, blue

[a] Dynamic Single-Mode (DSM) operation

oscillation. In the early 1960s, when the study of HeNe, ruby and other lasers was started, many laboratories or institutions also began research on semiconductor lasers. At that time, the Gunn effect capable of microwave oscillation was discovered and the related developments occured [1.8]. Furthermore, it was realized that a semiconductor laser must be considered not only as a diode, but also as an optical waveguide. The basic concept of a direct modulation limit [1.9] was discovered at this time.

The second stage of development began with the introduction of a DH and its successful continuous oscillation at room temperature [1.10, 11]. The development of reliable GaAlAs/GaAs semiconductor lasers continued and was detailed by *Hayashi* [1.12]. In 1975, well-balanced optical-fiber communication systems appeared in combination with a Si photodetector and optical fibers. In the meantime, the concept of transverse-mode control was clarified, various structures were proposed [1.13-42].

Then a low-loss band of a silica optical fiber was discovered in the 1 μm wavelength region [1.43]. While attending the International Conference at Nemunosato in Japan, *Hsieh* of MIT Lincoln Laboratory [1.44] presented a paper on a GaInAsP/InP laser proving a 1.1 μm-wavelength output.

The third stage began with GaInAsP/InP semiconductor lasers. An investigation of low-loss optical fibers, along with a reliability study of this system, was completed in a short time and greatly contributes to optical-fiber communications [1.45-81]. Concurrently, *Suematsu* of Tokyo Institute of Technology proposed the concept on Dynamic Single-Mode (DSM) lasers in 1980 [1.82], which led to the study of Distributed FeedBack (DFB) and Distributed Bragg Reflector (DBR) lasers [1.83-97].

Another important industrial field appeared in 1984: optical-disc devices such as Compact Discs (CDs) and Video Discs (VDs). Optical digital audio discs were developed and the annual production of laser chips soon reached 10 000 000 per year. We may say that the fourth stage started with high-performance lasers for optical communications, short-wavelength lasers for optical discs [1.98-117], and the demand for various functions and coherence controlling after 1985 [1.42, 118-121].

The 1990s found us entering the fifth generation of semiconductor lasers. Devices emitting in wider spectral ranges and at higher performances have been developed.

1.5 Typical Semiconductor Lasers

Semiconductor lasers are capable of producing a continuous output of greater than several tens of mW at the visible to infrared wavelengths. The most reliable device may have a lifetime of several 100 000 hours. Commercial GaAlAs lasers emitting at 0.78 μm provide an output of 10 mW and a guaranteed lifetime of 8000 h.

GaInAsP/InP semiconductor lasers are uitilized as light sources for optical communications. Typical wavelengths are 1.3, 1.48 and 1.55 μm. With the progress of quaternary-crystals growth techniques, integration of photo-electronic or photonic devices has become possible.

An InGaAs/GaAs strained-layer superlattice is employed for obtaining 0.98 μm wavelength output power to become a pumping source for erbium-doped optical fiber amplifiers [1.122, 123]. The compound semiconductor GaInAlP grown on a GaAs substrate has been developed for a visible laser emitting $0.63 \div 0.67$ μm light[1] [1.107-117] .

Green- and blue-light semiconductor lasers are difficult to fabricate due to the problem of p-type doping, but pulsed operation in the green to blue wavelength region using a ZnSSe system was achieved in 1991 [1.124] and a blue ZnMgSSe laser operating at 77 K was demonstrated in 1992 [1.125].

---

[1] The symbol $\div$ is used throughout the text as a shorthand for "from – to" or "between".

2. Materials for Semiconductor Lasers

Compound semiconductors, especially III-V compound semiconductors, are well suited for semiconductor lasers. The most basic, necessary condition required of laser materials is, of course, that the input energy can be converted into light energy with a reasonably high efficiency. In addition to this, the injected electron and hole concentrations should be higher than approximately $2 \cdot 10^{18}$ cm^{-3}, so that direct recombination can occur and optical gain is sufficient to reach the lasing threshold. Materials research is important, with the goal of finding materials capable for producing various semiconductor lasers with a wide range of lasing wavelengths. In Chap. 2 we concentrate on investigating material properties which determine the aforementioned basic characteristics of semiconductor lasers.

2.1 III-V Compound Semiconductors

2.1.1 Band Structure of III-V Semiconductors

The most important factor in semiconductor-laser materials is whether injected carriers can change into light. So-called *direct-band-gap semiconductors* can be used for light-source devices. Figure 2.1 [2.1] depicts the band structure of GaAs. The electron energy [eV] is plotted on the vertical axis and the crystal wave vector **k** on the horizontal axis. The band structure depends on the crystal direction. Figure 2.2 shows the first Brillouin zone of GaAs, with important lines and points depicted with Γ, L and X corresponding to the Γ, L and X points in Fig. 2.1 [2.2]. The center of the zone is denoted as Γ. The intersection of the [111] axes with the zone edge is L and the intersection of the [100] axes with the zone edge is X.[1]

Upper bands are conduction bands, while the lower bands are valence bands. In a direct-band-gap semiconductor, the minimum energy in the con-

---

[1] Several kinds of parentheses to designate crystal planes and directions are used in the conventional fashion [2.3]:

    (hkl)   Miller indices for a single plane.

    ⟨hkl⟩   Miller indices for a full set of equivalent direction.

    [hkl]   Miller indices for the direction of a crystal.

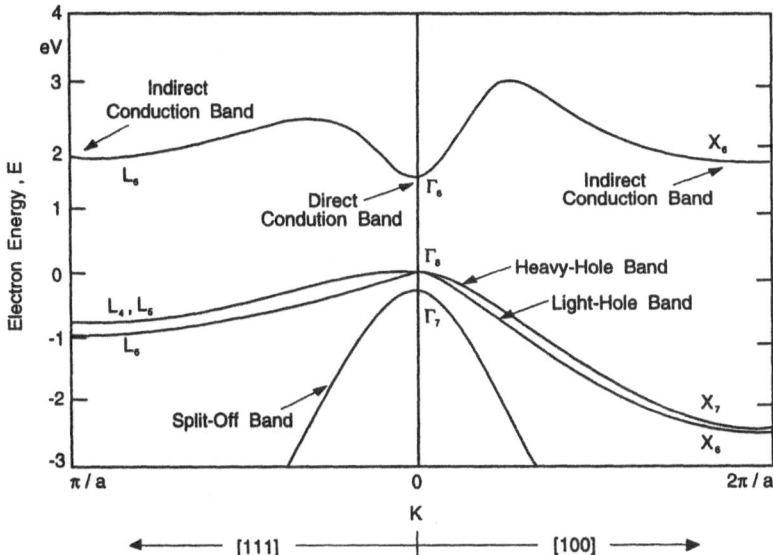

Fig.2.1. The band structure of GaAs with the energy E plotted as a function of momentum wave vector **k** along the [100] and [111] directions. The Γ, X and L designations identify the interband energy gaps [2.1]

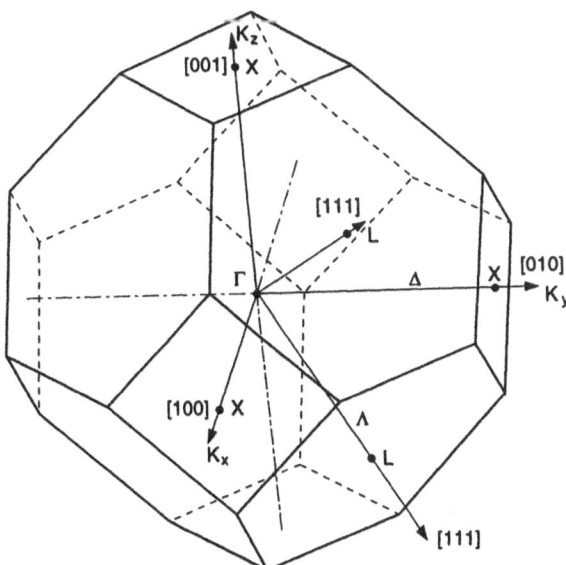

Fig.2.2. The first Brillouin zone of the diamond and zinc-blende lattices, with the most important symmetry points and lines indicated. The equivalence of some points are evident. For example, there are six equivalent X points [2.2]

duction band and the maximum energy in the valence band occur at the same value of **k**. On the other hand, in an indirect-band-gap semiconductor, this condition does not occur, for instance, when L_6 or X_6 are lower than Γ_6, as in Fig.2.1. In GaAs, the emission or the absorption of a photon therefore occurs at **k** = 0 between Γ_6 and Γ_8 whose energy gap is 1.424 eV.

Electrons and holes in crystals move like free particles with effective masses. The electron effective mass m_e^* and the hole effective mass m_h^* are given by

$$m_e^* = \left[\frac{\partial^2 E_c(k)}{\partial k^2} \right]_{k=0}^{-1} \quad \text{and} \quad m_h^* = -\left[\frac{\partial^2 E_v(k)}{\partial k^2} \right]_{k=0}^{-1} \qquad (2.1,2)$$

where $E_c(k)$ and $E_v(k)$ are the energies of the conduction-band edge and the valence-band edge, respectively. In GaAs, (2.1, 2) indicate that the effective mass depends on the curvature at Γ_6 or Γ_8. The smaller curvature implies a smaller effective mass leading to a higher mobility. Table 2.1 lists the energy gaps and the density-of-states effective masses for GaAs, AlAs and GaAlAs. The free-electron mass is denoted by m_0.

Fortunately, many direct-band-gap semiconductors are among III-V compound semiconductors. Additionally, a wide range of wavelengths can be covered by mainly III-V compound semiconductors. Figure 2.3 illustrates

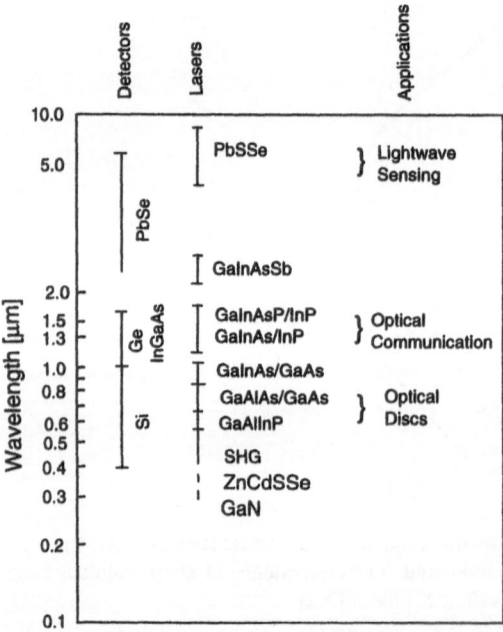

Fig.2.3. Wavelength range of compound semiconductor lasers and their field of application

Table 2.1. Parameters for GaAs, AlAs and $Al_xGa_{1-x}As$ [2.3]

GaAs	AlAs	$Al_xGa_{1-x}As$
E_g^Γ (direct) $= 1.424$ eV at 297 K	E_g^Γ (direct) $= 3.01$ eV	$E_g^\Gamma (0 < x < 0.45) = 1.424 + 1.247x$
E_g^L (indirect) $= 1.708$ eV at 297 K	E_g^L (indirect) $= 2.25 \div 2.35$ eV	$E_g^\Gamma (0.45 < x < 1.0) = 1.424 + 1.247x + 1.147(x-0.45)^2$
E_g^X (indirect) $= 1.900$ eV at 297 K	E_g^X (indirect) $= 2.168$ eV	$E_g^L = 1.708 + 0.642x$
		$E_g^X = 1.900 + 0.125x + 0.143x^2$
$m_p = 0.48m_0$	$m_p = 0.79m_0$	$m_p = (0.48+0.31x)m_0$
$m_n^\Gamma = 0.067m_0$	$m_n^\Gamma = 0.15m_0$	$m_n^\Gamma = (0.067+0.08x)m_0$
$m_n^L = 0.55m_0$, $m_n^X = 0.85m_0$	$m_n^L = 0.67m_0$	$m_n^L = (0.55+0.12x)m_0$
$m_0 = 9.11 \cdot 10^{-28}$ g	$m_n^X = 0.78m_0$	$m_n^X = (0.85-0.07x)m_0$
$E_g(T) = 1.519 - 5.405 \cdot 10^{-4} T^2/(204+T)$		

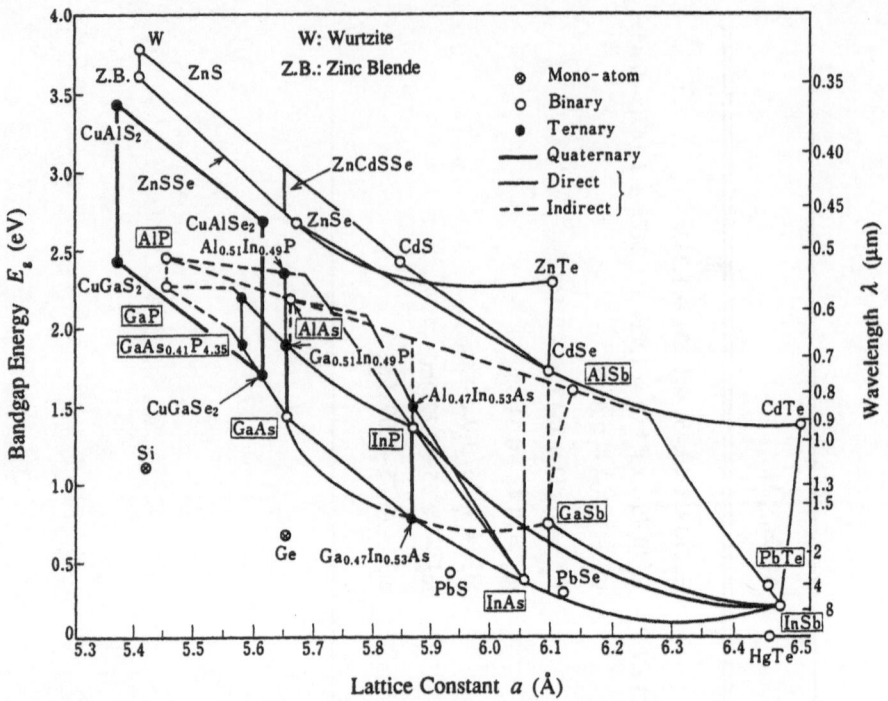

Fig.2.4. Band-gap energy versus lattice constant of several kinds of compound semi-conductors

the wavelength range of compound semiconductors and their applicable use, covering a range from 0.4 μm to 10 μm. It is also possible to grow semiconductors with various band-gap energies on a substrate. This enables the formation of a so-called *double heterostructure*, an important structure in which the active layer is sandwiched by wider-band-gap semiconductors (Sect. 1.3).

Figure 2.4 displays the band-gap energy vs. the lattice constant of several compound semiconductors. Binary crystals such as GaAs and InP are now popular as substrates because fewer atomic species lead to an easily controllable and highly qualitative crystal growth.

Figure 2.4 illustrates that it is possible to grow AlAs, $Ga_{1-x}Al_xAs$, $Al_{0.51}In_{0.49}P$, $Ga_{0.51}In_{0.49}P$ on a GaAs substrate. Ternary materials are composed of two kinds of binary materials. For example, $Ga_{1-x}Al_xAs$ consists of AlAs and GaAs at the rate of x and 1-x, respectively. The band-gap energy changes according to the rate of Al content x, as listed in Table 2.1. In many ternary crystals the lattice constant also changes according to the composition. In the $Ga_{1-x}Al_xAs$ system, the lattice constants of the binary crystals, GaAs and AlAs, happen to be close to each other. Thus, the lattice constant of $Ga_{1-x}Al_xAs$ is close to the GaAs substrate, too. From Vegard's law, the lattice constant can be approximated by assuming proportionality to the molecular percentage of compositions. The lattice constant of $Ga_{1-x}Al_xAs$ is therefore given by

$$a_{GaAlAs} = xa_{AlAs} + (1-x)a_{GaAs} .$$ (2.3)

The band-gap energy can also be described as

$$E_g = xE_{g,AlAs} + E_{g,GaAs} - K_{GaAlAs}x(1-x)$$ (2.4)

where K_{GaAlAs} is the bowing parameter. In quaternary crystals, such as the GaInAsP crystal, both the lattice constant and the band-gap energy can easily be changed to suit the requirements.

The relationship between wavelength and band-gap energy is as follows:

$$\lambda\ [\mu m] = \frac{1.2398}{E_g\ [eV]} .$$ (2.5)

Consequently, the band-gap energy can be obtained with (2.5) during the initial design of a semiconductor laser. The composition can then be decided from both the lattice constant and the band-gap energy.

2.1.2 Other Characteristics of III-V Compound Semiconductors

The crystal structure of III-V compound semiconductors is a zinc-blende (or wurtzite) structure. Figure 2.5 exhibits the zinc-blende crystal structure of GaAs or InP. Because the minimum number of bonds occur at the PQ facet, the material is cleaved along the PQ facet. The PQ facet is called a *(111) facet*. The facet consisting of only group-III atoms (Ga, In) is termed a *(111) A facet*. The facet containing only group-V atoms (As, P) is named a *(111) B facet*. Although the (111) A facet does not have unpaired electrons, the (111) B facet does have them. Thus, the (111) B facet is chemically active.

Thermal conductivity decreases with the increase of disorder because the lattice obstructs the flow of thermal vibration. For instance, the thermal conductivity of $Ga_{1-x}Al_xAs$ decreases as the increase of Al content x from x = 0 to 0.5, as shown in Fig.2.6 [2.4, 5]. The Al atom replaces the Ga atom of GaAs at the rate of x thus preventing thermal flow. The thermal conductivity reaches a minimum near x = 0.5. The increase of x from 0.5 to 1 then leads to an increase of the thermal conductivity.

Generally speaking, the electron mobility of $Ga_{1-x}Al_xAs$ is lower than that of GaAs due to the alloy scattering caused by the disorder of Ga and Al atoms in group-III atomic sites. The lower-band-gap materials generally

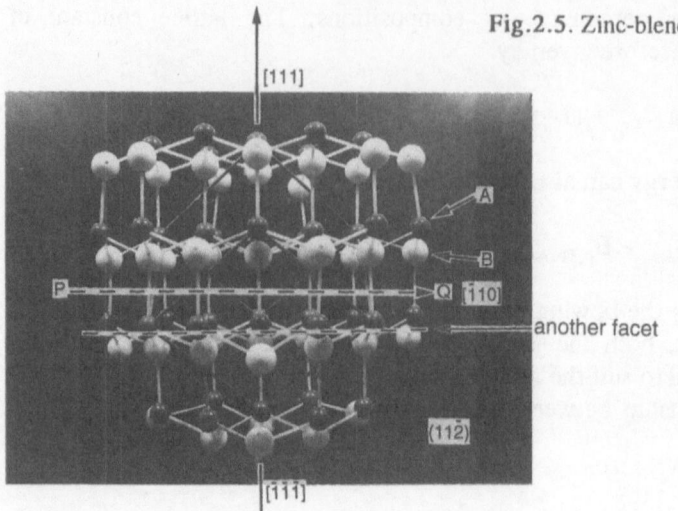

Fig.2.5. Zinc-blende crystal structure

Zinc-blende Crystal Structure
A: III Group Atoms   B: V Group Atoms

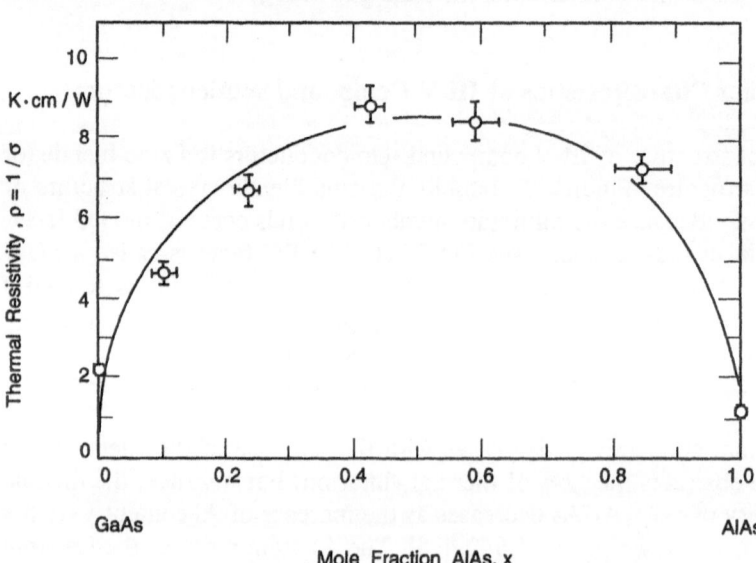

Fig.2.6. Thermal resistivity as a function of the AlAs mole fraction [2.5]

have a higher mobility. The electron mobility of InSb is $\approx 100\,000$ cm^2/ V·s, which is about 100 times faster than that of Si.

2.2 Crystals for Visible to Near-Infrared-Wavelength Emission Semiconductor Lasers

2.2.1 Importance of Visible to Near-Infrared-Wavelenth Laser Emission

The importance of the visible to near-infrared-wavelength region, i.e., the short-wavelength region, is characterized as follows [2.6]:

Diffraction Limit of Light. The diameter of the focused spot is given by

$$\Delta D = 1.22\lambda/NA \qquad (2.6)$$

where λ denotes the wavelength, and NA is the Numerical Aperture of a focusing lens. For example, ΔD is equal to 1.9 μm when the wavelength is 0.78 μm and NA $= 0.5$.

Sensitivity of Photo-Sensitive Materials. The sensitivity of photo-sensitive materials to light becomes high for $\lambda < 0.7$ μm, thus making a laser with a short wavelength desirable.

Visual Effect. Human eyes are sensitive in the wavelength range of $0.4 < \lambda < 0.8$ [μm], being one octave. The eye's sensitivity peaks at 0.55 μm.

Color Display. Lasers containing the three primary colors, red (R), green (G) and blue (B), are utilized for displays [2.7].

2.2.2 Crystal Materials for the Near-Infrared Region

The GaAlAs/GaAs system is now at present the most popular material for the near-infrared region. GaAlAs/GaAs lasers are used for audio and visual discs, short-span optical-fiber communication, opto-electronics and integrated circuits, and so forth. One of the most attractive features is that high-speed electron devices can be combined with optical devices monolithically in the GaAlAs/GaAs system. The first laser oscillation [2.8-10] and the first CW operation were observed in such a system [2.11]. Additionally, achieving an optical-fiber loss of 20 dB/km [2.13], along with the success of the CW operation of the GaAlAs/GaAs laser in 1970, acted as a spur to opto-electronics research.

Fortunately, the lattice constant of $Ga_{1-x}Al_xAs$ (x = 0 to 1) is almost the same as that of GaAs. Binary lattice constants of the system at 0°C are 5.65325 Å for GaAs and 5.6605 Å for AlAs, as shown in Fig.2.4. As previously mentioned, changing the energy gap while keeping the same lattice

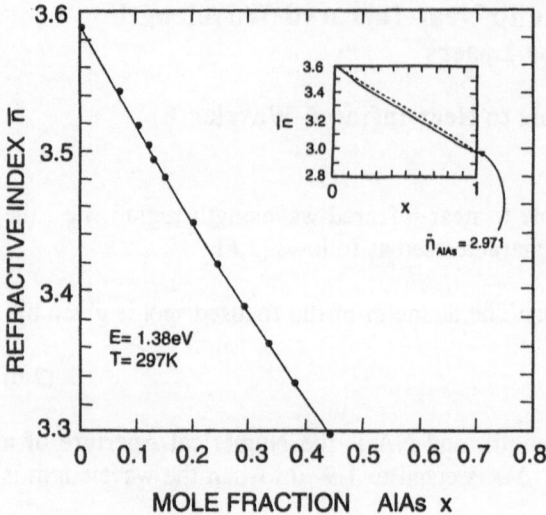

Fig.2.7. Refractive index of AlGaAs as a function of x at the heterostructure-laser emission energy of 1.38 eV. The dashed line in the insert is a linear extrapolation of the refractive index of 3.590 for GaAs to the refractive index of 2.971 for AlAs [2.14]. The uncertainty in the refractive index values for these measurements is ±0.005, while the uncertainty for the technique used to assign x is generally taken as ±0.02 [2.13]

constant is possible in ternary materials. The largest lattice mismatch $\Delta a/a$ of the system, which occurs between GaAs and AlAs, is less tha 0.13%. The necessary condition for epitaxial growth is that the lattice mismatch should be less than 0.1%. Thus, we can fabricate a $Ga_{1-x}Al_xAs$ epitaxial layer on GaAs substrate.

The energy gap varies with the Al content x as

$$E_g (0 < x < 0.45) = 1.424 + 1.247x .\tag{2.7}$$

The transition mechanism changes from a direct transition to an indirect one when the Al content x exceeds 0.45. From (2.5) we can expect a wavelength to range from 0.87 to 0.62 μm. However, approximately 0.67 μm

Table 2.2. Refractive Index of $Al_xGa_{1-x}As$ at 297 K for E = 1.38 eV [2.4] listed for different x values

x: 0	0.05	0.10	0.15	0.20	0.25	0.30	0.35	0.40	0.45	0.50	0.55	0.60	0.65	0.70
n: 3.590	3.555	3.520	3.486	3.452	3.418	3.385	3.353	3.321	3.289	3.258	3.227	3.197	3.167	3.138

would be the shortest wavelength, as stated in Sect.2.2.3. The refractive index also changes with x, as shown in Fig.2.7 and summarized in Table 2.2. The refractive index of GaAlAs for a photon energy of 1.38 eV can be expressed as

$$n(x) = 3.590 - 0.710x + 0.091x^2 . \tag{2.8}$$

2.2.3 Crystal Materials for Visible Laser Emission

In order to utilize lasers for opto-electronic devices, visible lasers with short wavelengths are very convenient. If the Al content x of the active layer of a $Ga_{1-x}Al_xAs$ system is increased, oscillations with short wavelengths are obtained. However, if x exceeds 0.4, the minimum value of the conduction band transfers from the Γ valley to the X valley, as illustrated in Fig.2.8. For this reason, the threshold current density starts rising around 0.74 μm. Therefore, in this system the wavelength of 0.67 μm will become a limit.

In the shorter-wavelength region a Double-Hetero (DH) structure of the quaternary $Ga_{1-x}In_xAs_yP_{1-y}$ which uses GaAs as a substrate can be utilized. Formation of a cladding layer by $Ga_{1-x}Al_xAs$ is being studied.

Making the lattice constant of a substrate small enables short-wavelength (wavelengths of 0.7μm or less) laser emission. Figure 2.4 displays a lattice constant and a band gap.

As for an injection-type laser with a large band-gap energy, it was impossible to grow a quaternary InGaAlP crystal on a GaAs substrate by the Liquid Phase Epitaxy (LPE) method. Using the vapor-phase growth method, a laser in the 0.63 μm region has become possible.

Among III-V crystals, 0.5 μm is the short-wavelength limit. A GaP green-emitting diode is of the indirect transition-type with a small recombination rate of electrons and holes; and laser oscillation is not possible. In

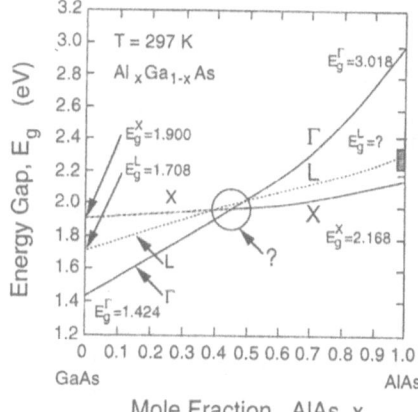

Fig.2.8. The compositional dependence of the direct $\Gamma_8 \to \Gamma_6$ and the indirect $\Gamma_8 \to X_6$ and $\Gamma_8 \to L_6$ energy gaps for $Al_xGa_{1-x}As$. [2.4]

other words, for green and blue lasers, II-VI semiconductors such as ZnCdSSe/ZnSe [2.15], ZnMgSSe/ZnSe [2.16], etc. have larger band gaps than III-V semiconductors, and emit strong luminescence at 0.45 to 0.35 μm. In 1990, a p-type II-VI semiconductor ZnSe was obtained by N_2 doping to make a p-n junction, creating a ZnCdSSe/ZnSe green laser under pulsed condition at room temperature.

One of the principal features of a semiconductor laser is that it starts oscillating when a low voltage of several volts is applied [2.17]. Thus, a p-n junction is essential. A technique that may enable p-n junctions to be formed in II-VI semiconductors is to introduce group-I atoms to control the impurities. An further useful technique involves compound crystals to match the lattice constants. By utilizing a nonlinear effect, blue light of 0.4 μm or shorter can be obtained from the second harmonic of a semiconductor-laser beam operating in 0.8 μm band.

2.3 Crystals for Semiconductor Lasers with 1-μm and Longer Emission Wavelengths

2.3.1 Importance of the 1-μm Emission Wavelength

Due to the development of optical-fiber communications, GaAlAs lasers (0.85 μm band) have been studied since 1970. Devices with higher reliability were introduced and the so-called *first-era lightwave system* was realized [2.18].

On the basis of optical-fiber-loss measurement *Kao* et al. [2.19] predicted that long wavelengths would have a smaller loss if the water absorption was removed from a silica fiber. Their prediction generated optical communication in the long-wavelength region. A new method to produce a silica fiber was developed in 1974, that is, Modified Chemical Vapor Deposition (MCVD). In 1976, a joint research effort between NTT and Fujikura Cable Co. produced a record-breaking fiber with a transmission loss of 0.6 dB/km at wavelengths of 1.3 μm [2.20]. This accomplishment attracted attention to research in this wavelength region. During the second era, a transmission loss of 1 dB/km or less (the minimum of 0.154 dB/km at 1.55 μm) was achieved [2.21].

Since 1978, materials for GaInAsP with longer wavelengths of 1.2 $\div$ 1.6 μm have been studied. Characteristics of long-wavelength fiber communications in the second era can be summarized as follows:

● Capabilities for achieving the lowest transmission loss and a longer relay distance are possible. Figure 2.9 depicts the transmission loss and the dispersion of an optical fiber. At the end of 1978, NTT first achieved a 0.2

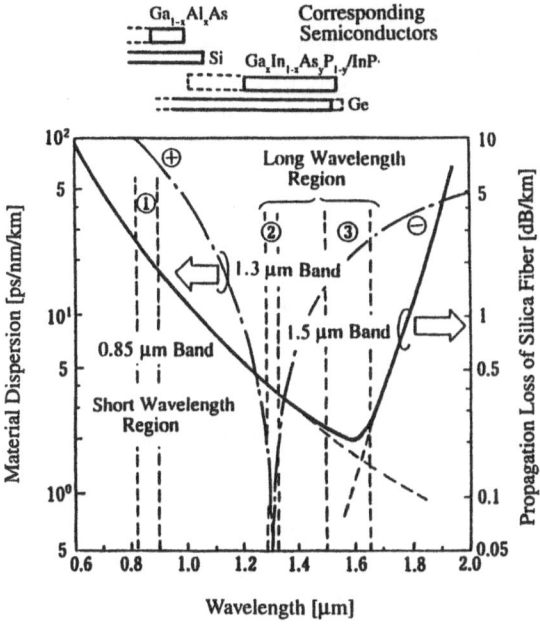

Fig.2.9. Transmission loss of a silica fiber

dB/km in the 1.5-μm wavelength region [2.21] and then surpassed it by obtaining the lowest limit (0.154dB/km) later [2.22]

● A wavelength-multiplexing system which transmits a number of wavelengths to a single fiber can be utilized due to the transmission loss dropping to 1 dB/km or less in the wavelength region of 1.1 to 1.6 μm.

● The availability of a relatively large diameter of the fiber core, satisfying the single-mode transmission condition of optical fibers.

● Rayleigh scattering, one of the causes of light loss in lasers and optical circuits, becomes small in the long-wavelength region, with scattering being proportional to λ^{-4}.

● The small light absorption loss in a Faraday rotator. (It is convenient to consider light absorption loss for a polarization rotation as a new figure of merit).

● The material dispersion which controls the transmission band of an optical fiber is within ± 3 ps/km·Å (distance)·(linewidth).

The transmission loss of silica fiber becomes small in the wavelength region of 1.55 μm, which is advantageous for long-distance optical communications such as undersea cables. However, since the dispersion value is large, the linewidth of the employed light sources must be very small, requiring the use of a dynamic single-mode laser. Optical communication as described above will be denoted as the *communication of the third era.*

As the wavelength becomes longer (Fig.2.9), the transmission loss increases due to intrinsic infrared absorption. Extensive studies on new optical fibers predict the possibility in the near future of optical communication in the longer-wavelength region of λ = 2 to 10 μm. With increasing wavelength, the photon energy $h\nu$ (h: Planck's constant, ν: frequency) becomes small; that is, $h\nu$ = 0.12 eV at 10 μm. The thermal energy $k_0 T$ (k_0: Boltzmann constant, T: absolute temperature) is 0.03 eV at 70°C, which cannot be ignored when compared with the photon energy at long wavelengths. Therefore, coherent optical communication takes essentially place in the region of λ > 2 μm, and is referred to as the *fourth* or *fifth era of optical communications*.

2.3.2 Crystal Materials for the 1-μm Emission Wavelength

Ternary or quaternary semiconductor crystals are used since binary semiconductor crystals with 1-μm band gaps are not available. Matching the lattice constants is important for semiconductor devices such as semiconductor lasers, LEDs with a current density of 5 kA/cm^2 · μm or a high-output density of 1 mW/μm^2, or photodiodes for low-noise photodetectors.

The band gap of a quaternary crystal can change widely while maintaining the lattice match completely to a binary crystal used as a substrate. Figure 2.4 depicts the relationship between the lattice constant and the band gap of binary, ternary and quaternary crystals. An example is $Ga_x In_{1-x}$ · $As_y P_{1-y}$, which utilizes InP (a = 5.8696 Å) as a substrate, where the band gap can be changed in the region of $0.7 \leq E_g \leq 1.35$ [eV] when the composition is changed along the vertical line. The corresponding emission wavelength ranges from 0.92 to 1.67 μm. The ternary materials with lattice constants matching the InP substrate are $Al_{0.47} In_{0.53} As$ and $Ga_{0.47} In_{0.53} As$. The relationships between x, y and the band-gap energy of $Ga_x In_{1-x}$ · $As_y P_{1-y}$, the lattice constant which is matched to InP, are

$$x = \frac{0.466y}{1.03 - 0.03y} \quad (0 \leq x \leq 1), \tag{2.10}$$

$$E_g(y) = 1.35 - 0.72y + 0.12y^2 \quad [eV] . \tag{2.11}$$

No longer are x and y independent of one another because the lattice constant is adjusted to 5.86875 Å. Consequently, the band-gap energy is expressed only by the Ga or As contents. The band-structure parameters of GaInAsP/InP are summarized in Table 2.3.

Table 2.3. Band-structure parameters of $Ga_xIn_{1-x}As_yP_{1-y}$ lattice matched to InP

Parameter	Dependence on the mole fractions x and y
Energy gap at zero doping	E_g [eV] $= 1.35 - 0.72y + 0.12y^2$
Heavy-hole mass	$m_{hh}^*/m_0 = (1-y)[0.79x + 0.45(1-x)] + y[0.45x + 0.4(1-x)]$
Light-hole mass	$m_{lh}^*/m_0 = (1-y)[0.14x + 0.12(1-x)] + y[0.082x + 0.0261(1-x)]$
Dielectric constant	$\epsilon = (1-y)[8.4x + 9.6(1-x)] + y[13.1x + 12.2(1-x)]$
Spin-orbit splitting	Δ [eV] $= 0.11 + 0.31y - 0.09x^2$
Conduction-band mass	$m_c^*/m_0 = 0.080 - 0.039y$

Compound crystals corresponding to the 1-μm laser-emission wavelength region are as follows:

(1) $Ga_xIn_{1-x}As_yP_{1-y}$ (InP): $0.92 < \lambda_g < 1.67$ [μm] ,

(2) $(Ga_{1-x}Al_x)_yIn_{1-y}As$ (InP): $0.83 < \lambda_g < 1.55$ [μm] ,

(3) $Ga_{1-x}Al_xAs_ySb_{1-y}$ (GaSb): $0.8 \ \ < \lambda_g < 1.7$ [μm] ,

(4) $Ga_xIn_{1-x}As_ySb_{1-y}$ (InAs): $1.68 < \lambda_g < 2$ [μm] ,

(5) $Ga_xIn_{1-x}As_ySb_{1-y}$ (GaSb): $1.8 \ \ < \lambda_g < 2$ [μm] .

The substrate is given in parentheses. Crystal growth of these materials is possible with a lattice mismatch of $\pm 0.1\%$ or less. The $Ga_xIn_{1-x}As_yP_{1-x}$ and InP heterojunction is mainly studied as a material system for lasers or photo diodes.

2.3.3 Longer-Wavelength Materials

Fluoride-glass fibers are promising for long-distance optical communication in the $2 \div 4$ μm wavelength range. The loss in fluoride-glass fiber is predicted to be one or two orders of magnitude lower than that of silica fiber. In addition, this wavelengths band is important for lidar and optical remote sensing [2.24]. Thus, a potential materials system to cover the wavelength range from 1.7 to 5 μm is GaInAsSb/AlGaAsSb [2.25].

3. Basic Design of Semiconductor Lasers

A double heterostructure is a very basic, necessary design for semiconductor lasers capable of confining both electronic carriers and lightwaves in its cavity. The introduction of double heterostructures into a GaAlAs/GaAs laser in 1970 provided a breakthrough in obtaining continuous operation at room temperature. Owing to the importance of double heterostructures, we explain the functions and design of a semiconductor laser cavity.

3.1 Double Heterostructures and Their Design

3.1.1 Double Heterostructures

A Double Heterostructure (DH) consists of a light amplification layer and cladding layers, a so-called *active layer sandwiched between two cladding layers with a larger band-gap energy*. The cladding layers are doped for current injection. When a forward bias is applied, a number of electrons are injected from the n-type cladding layer into the active layer. Holes are supplied from the p-type cladding layer and both carriers are confined in the active layer because the hetero-barriers of the conduction and valence bands inhibit carrier flow. The active layer has a refractive index larger than those of the cladding layers. Light can then be confined in the active layer which behaves as a dielectric waveguide with the active layer functioning as its core. Thus, both carriers and the optical mode in the active layer of a double heterostructure should be confined. A significant improvement brought about by the introduction of double heterostructures is the considerable reduction of the threshold current required for lasing operation to begin.

A DH semiconductor laser produced with GaAlAs/GaAs is illustrated in Fig.3.1. Layers 2 to 5 are successively grown on the wafer (region 1). The GaAs is the active layer for laser oscillation, while the junction between the layers 2 and 3 (or 3 and 4) is the p-n heterostructure junction. Region 5 is the highly doped cap layer where the ohmic electrode is at-

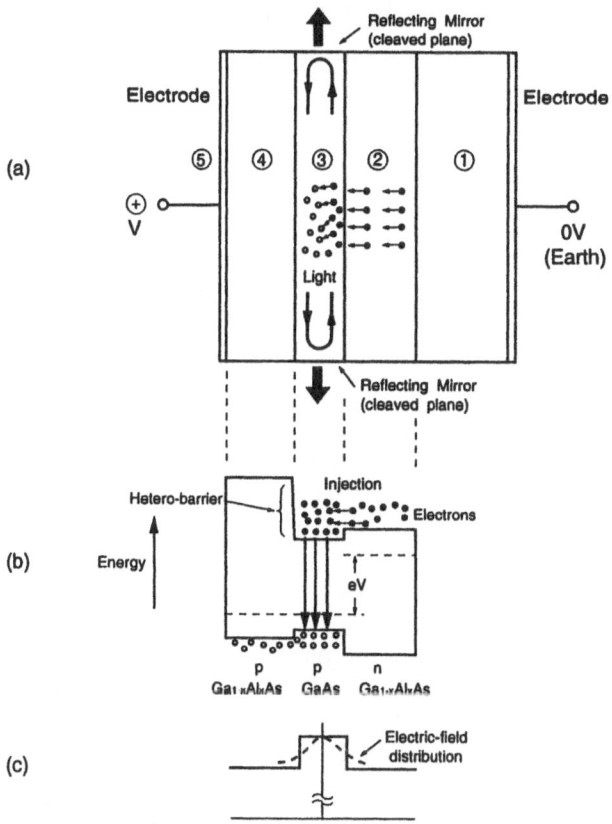

Transmitted Light

Reflecting Mirror
(cleaved plane)

Electrode

Electrode

(a)

⑤ ④ ③ ② ①

(+)
V

0V
(Earth)

Light

Reflecting Mirror
(cleaved plane)

Injection

Hetero-barrier

Electrons

(b) Energy

eV

p   p   n
Ga₁₋ₓAlₓAs GaAs Ga₁₋ₓAlₓAs

Electric-field
distribution

(c)

Fig.3.1. A double-heterostructure semiconductor laser: (**a**) Layer structure, (**b**) energy band, and (**c**) refractive-index and electric-field distributions [3.1,2], (**d**) SEM photograph of a GaInAsP/InP DH laser

tached to. When a voltage V approximately equal to the energy gap of the GaAs active layer (3 in Fig.3.1b) is applied, electrons and holes are injected into the active layer.

3.1.2 Design of Double-Heterostructure Lasers

In Sects.3.1.2-4 we explain how to design DH lasers based on both carrier and optical-confinement considerations. The GaAlAs/GaAs system is used as an example; and the GaInAsP/InP system is designed in a similar manner.

The band gap of $Ga_{1-x}Al_xAs$ material allows lasing wavelengths ranging from 0.67 to 0.87 μm. The hetero-barrier height necessary for DH lasers is usually 0.3 eV. For instance, if the material has been designed to give a lasing wavelength of 0.78 μm, the corresponding band gap of the active layer ($E_{g,act} = 1.589eV$), which is determined from

$$E_g \text{ [eV]} = \frac{1.2398}{\lambda \text{ [}\mu\text{m]}} . \tag{3.1}$$

The Al content of the active layer is therefore $x_{Al} = 0.13$ from

$$x_{Al} = \frac{E_g - 1.424}{1.247} \quad (E_g < 1.984eV),$$

$$x_{Al} = \frac{-0.2147 + [0.046 - 4.558(1.656 - E_g)]}{2.294} \quad (1.985eV < E_g). \tag{3.2}$$

Currently the band-gap energy and the Al content of cladding layers are given as $E_{g,clad} = E_{g,act} + 0.3$ eV $= 1.889$ eV, and therefore $x_{Al} = 0.37$.

3.1.3 Energy-Band Diagram of DH Lasers

Figure 3.2 depicts the various structures of junctions in semiconductors with different band gaps. Let us discuss the junction region where mainly a p-n or n-p junction is fabricated. The active layer of a semiconductor laser usually has a narrow band gap and if two of the four, including a p-n or n-p junction, in Figs.3.2b-e are combined, a double-hetero junction can be observed.

The work function and the band gap of a narrow-gap and a wide-gap semiconductor are assumed to be χ_1 and χ_2, and E_{g1} and E_{g2}, respectively; other relevant symbols are listed in Fig.3.2a. When two semiconductors with different Fermi levels are joined together, carrier transport between them occurs until the Fermi levels coincide and reach thermal equilibrium. When equilibrium has been achieved, the built-in potential V_D is applied across the p-n (or n-p) junction prohibiting carrier overflow. The built-in potential is described as

$$V_D = \frac{F_2 - F_1}{e} = \frac{E_{g1} + \Delta E_c - \delta_1 - \delta_2}{e}. \tag{3.3}$$

When the p-p junction is formed, the energy state changes, as shown in Fig.3.2b. In other words, the straight-line Fermi level produced above (near) the valence band is owing to carrier transport between semiconductors and a slight difference in the valence-band edge, and a large difference in the conduction-band edge are generated. The difference between band-

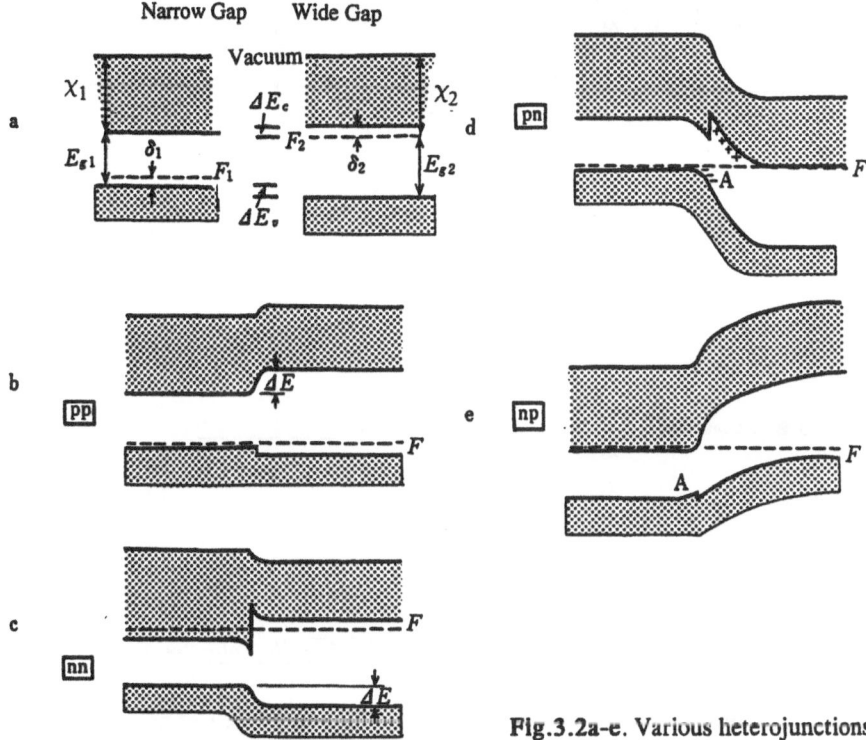

Fig.3.2a-e. Various heterojunctions

gap energies is expressed as $\Delta E_g = E_{g_2} - E_{g_1}$. Figure 3.2c illustrates a band structure where both layers are n-doped. In this case the straight-line Fermi level is produced near the conduction-band edge, and the energy difference is also expressed as $\Delta E_g = E_{g_2} - E_{g_1}$.

Next, consider the case where a p-n or n-p junction is formed, as illustrated in Fig.3.2d,e. The band structure will be detailed referring to *Anderson's* analysis and results [3.3]. According to *Dingle* [3.4], the ΔE_g of the GaAlAs/GaAs system is distributed to the hetero-barriers in the conduction band ΔE_c and valence band ΔE_v in the ratio of

$$\frac{\Delta E_c}{E_{g_2} - E_{g_1}} : \frac{\Delta E_v}{E_{g_2} - E_{g_1}} = 0.85 \pm 0.03 : 0.15 \pm 0.03 \; (X_{Al} < 0.4) \; . \; (3.4)$$

Recently, another experiment [3.5, 6] lead to the correction

$$\frac{\Delta E_c}{E_{g_2} - E_{g_1}} : \frac{\Delta E_v}{E_{g_2} - E_{g_1}} = 0.67 \pm 0.01 : 0.33 \pm 0.01 \qquad (3.5)$$

or

$$= 0.57 : 0.43 \; . \qquad (3.6)$$

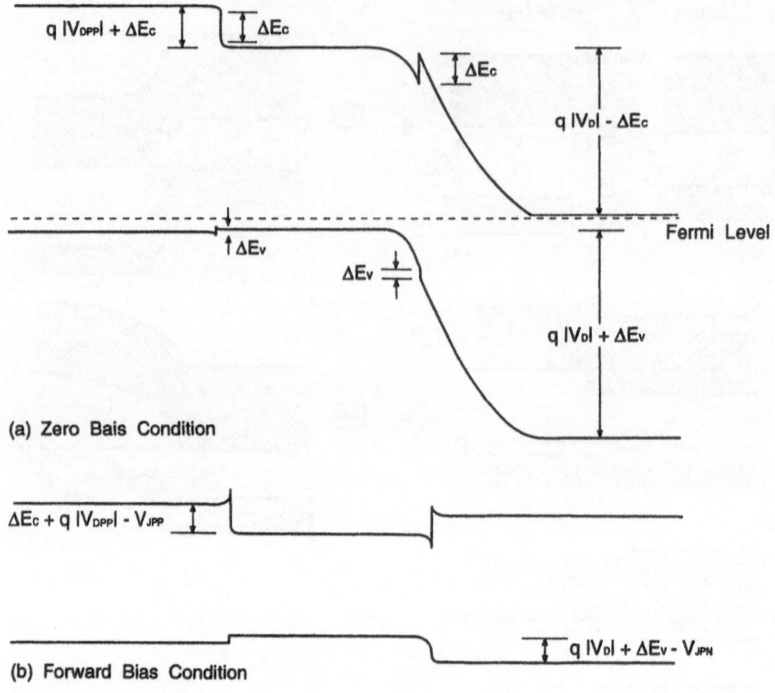

q |V$_{DPP}$| + ΔE$_c$ ΔE$_c$

ΔE$_c$

q |V$_D$| - ΔE$_c$

Fermi Level

ΔE$_v$

ΔE$_v$

q |V$_D$| + ΔE$_v$

(a) Zero Bais Condition

ΔE$_c$ + q |V$_{DPP}$| - V$_{JPP}$

q |V$_D$| + ΔE$_v$ - V$_{JPN}$

(b) Forward Bias Condition

Fig.3.3a,b. Energy-band diagram of p-type-Ga$_{0.7}$Al$_{0.3}$As-p-type-GaAs/ n-type-Ga$_{0.7}$Al$_{0.3}$As

In the present book (3.4) is employed.

For the GaInAsP/InP system, hetero-barriers are commonly expressed as [3.7]

$$\frac{\Delta E_c}{E_{g_2} - E_{g_1}} : \frac{\Delta E_v}{E_{g_2} - E_{g_1}} = 0.39 \pm 0.01 : 0.61 \pm 0.01 . \tag{3.7}$$

If the doping concentrations of a n-type cladding, p-type active and p-type cladding layers are given, the energy-band diagram can be drawn. Figure 3.3 depicts the energy-band diagram of a p-type GaAlAs/p-type GaAs/n-type GaAlAs DH structure. Hetero-barriers ΔE_c and ΔE_v, whose ratios are based on (3.4) occur at both p-p and p-n junctions. The difference between the Fermi levels is so small at the p-p junction that a small built-in potential occurs there, and at the same time a large built-in potential arises at the p-n junction. Thus, the energy-band diagram can be roughly drawn as in Fig. 3.3a. If a forward voltage V_a is applied to the double heterostructure, most of the voltage V_a is applied to the p-n junction. The n-type GaAlAs cladding layer is lifted against the p-type GaAs layers, as indicated in Fig.3.3b.

Next, the energy-band diagram is obtained by using the following model: Region 0: p-$Ga_{0.7}Al_{0.3}As$ ($N_A = 2 \cdot 10^{17} cm^{-3}$) Region 1: p-GaAs ($N_A = 5 \cdot 10^{17} cm^{-3}$) Region 2: n-GaAlAs ($N_D = 2 \cdot 10^{17} cm^{-3}$. The employed equations agree with those in the book of *Casey* and *Panish* [3.8]. First, focus on the p-GaAs/n-GaAlAs junction.

The procedure for drawing the band diagram is described as follows:

(i) Draw a horizontal line corresponding to the Fermi level.

(ii) Determine E_{g_1}, E_{g_2}, ΔE_c and ΔE_v. $E_{g_1} = 1.424$ eV, $E_{g_2} = 1.798$ eV, $\Delta E_c = 0.318$ eV and $\Delta E_v = 0.056$ eV from (3.1, 2 and 4).

(iii) Determine δ_1 and δ_2.

Next, we determine the positions of the valence-band edge E_v and conduction-band edge E_c from the Fermi level. They are obtained as follows:

$$N_D \simeq n = N_c \exp[(F-E_c)/kT] \quad \text{for} \quad N_D/N_c < 0.1, \tag{3.8}$$

$$N_A \simeq p = N_v \exp[(E_v-F)/kT] \quad \text{for} \quad N_A/N_v < 0.1 \tag{3.9}$$

where N_D and N_A are the donor and acceptor concentrations. N_c and N_v are the conduction-band and the valence-band effective densities of state, namely

$$N_c = 2\left[\frac{2\pi kT}{h^2}\right]^{3/2}\left[\left(\frac{m_n^\Gamma}{m_0}\right)^{3/2} + \left(\frac{m_n^L}{m_0}\right)^{3/2}\exp\left(-\frac{\Delta E^{L-\Gamma}}{kT}\right)\right.$$
$$\left. + \left(\frac{m_n^X}{m_0}\right)^3\exp\left(-\frac{\Delta E^{X-\Gamma}}{kT}\right)\right], \tag{3.10}$$

$$N_v = 2\left[\frac{2\pi kT}{h^2}\right]^{3/2}\left[(m_{ph})^{3/2} + (m_{pl})^{3/2}\right] \tag{3.11}$$

where h is Planck's constant, and m_0, m_n^Γ, m_n^L and m_n^X are the free electron mass, the density-of-states effective mases for the conduction band at Γ, L and X. m_{ph} and m_{pl} are effective masses of the heavy- and light-hole bands. $\Delta E^{L-\Gamma}$ and $\Delta E^{X-\Gamma}$ are $E_g^L - E_g^\Gamma$ and $E_g^X - E_g^\Gamma$, respectively. Equations (3.8, 9) imply that all of the donors or acceptors are ionized while the electron excitation from the valence band to the conduction band is negligible. Consequently, the carrier concentration is assumed to be equal to the doping concentration N_A and N_D.

Table 3.1 gives the actual data for calculating the bands. The value of $\exp(-\Delta E^{L-\Gamma}/kT)$ is $1.5 \cdot 10^{-5}$ in GaAs and those of between indirect bands are negligible. Then N_c and N_v for GaAs ($m_n^\Gamma = 0.067$, $m_p = 0.48m_0$) are described as

Table 3.1. Parameters for drawing a p-n junction (χ_1 and χ_2 denote work functions)

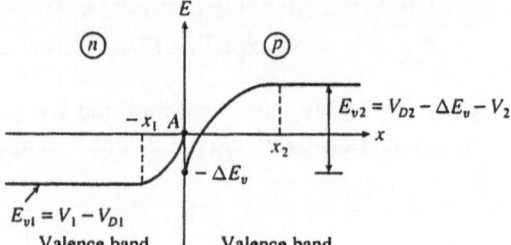

$$\Delta E_c = \chi_1 - \chi_2$$
$$\Delta E_v = (\chi_1 + E_{g2}) - (\chi_2 + E_{g1})$$

(pn) $$V_D = E_{g1} + \Delta E_c - \delta_1 - \delta_2$$
$$V_{D1} = V_D/K \quad V_{D2} = V_D - V_{D1}$$
$$V_1 = V_a/K \quad V_2 = V_a - V_1$$

$$K = 1 + \frac{\varepsilon_1(N_{A1}^- - N_{D1}^+)}{\varepsilon_2(N_{D2}^+ - N_{A2}^-)}$$

(pn) $$x_1^2 = \frac{2\varepsilon_1(V_{D1} - V_1)}{q(N_{A1}^- - N_{D1}^+)}$$
$$x_2^2 = \frac{2\varepsilon_2(V_{D2} - V_2)}{q(N_{D2}^+ - N_{A2}^-)}$$

(np) $$x_1^2 = \frac{2\varepsilon_1(V_{D1} - V_1)}{q(N_{D1}^+ - N_{A1}^-)}$$
$$x_2^2 = \frac{2\varepsilon_2(V_{D2} - V_2)}{q(N_{A2}^- - N_{D2}^+)}$$

δ_1: Fermi level measured from valence band of semiconductor 1
δ_2: Fermi level measured from conduction band of semiconductor 2

(a) *pn* junction

(b) *np* junction

$$N_c = 2.5 \cdot 10^{19} \left[\frac{m_n^\Gamma}{m_0}\right]^{3/2} \left[\frac{T}{300}\right]^{3/2} \text{cm}^{-3} \times 4.27 \cdot 10^{17}\,\text{cm}^{-3} \quad (297\text{K}), \quad (3.12)$$

$$N_v = 2.5 \cdot 10^{19} \left[\frac{m_p}{m_0}\right]^{3/2} \left[\frac{T}{300}\right]^{3/2} \text{cm}^{-3} \times 8.19 \cdot 10^{18}\,\text{cm}^{-3} \quad (297\text{K}) . \quad (3.13)$$

The parameters for $Ga_{0.7}Al_{0.3}As$ are as follows: $m_n^\Gamma = 0.092m_0$, $m_n^L = 0.59m_0$, $m_n^X = 0.83m_0$, $m_p = 0.57m_0$, $\Delta E^{L-\Gamma} = 0.103$ eV and $\Delta E^{X-L} = 0.152$ eV; therefore, from (3.10)

$$N_c = 9.3 \cdot 10^{17} \text{ cm}^{-3} \quad (297\text{K}) \tag{3.14}$$
$$N_v = 1.1 \cdot 10^{19} \text{ cm}^{-3} \quad (297\text{K}) \tag{3.15}$$

From (3.8, 9)

$$\delta_1 = F - E_v = -kT\ln(p/N_v), \tag{3.16}$$
$$\delta_2 = E_c - F = -kT\ln(n/N_c) . \tag{3.17}$$

The doping concentration for some semiconductor lasers exceeds $0.1N_c$ or $0.1N_v$. In these cases, higher-order approximations are required. The carrier concentrations are

$$n = N_c \, Fer_{1/2} \, [(F - E_c)/kT] \, , \tag{3.18}$$

$$p = N_v \, Fer_{1/2} \, (F/kT) \, . \tag{3.19}$$

Here, $Fer_{1/2}$ is the Fermi-Dirac integral

$$Fer_{1/2}(\xi) = 2\sqrt{\pi} \int_0^\infty \frac{\sqrt{\epsilon}\,d\epsilon}{1 + \exp(\epsilon - \xi)} \, . \tag{3.20}$$

The approximation of (3.18) by *Joyce* and *Dixon* [3.9] gives for $F - E_c$

$$F - E_c = kT \, [\ln(n/N_c) + 3.53553 \cdot 10^{-1} (n/N_c) - 4.95009 \cdot 10^{-3} (n/N_c)^2$$
$$+ 1.48386 \cdot 10^{-4} (n/N_c)^3 - 4.42563 \cdot 10^{-6} (n/N_c)^4] \, . \tag{3.21}$$

Generally, only a few terms of (3.21) are enough to calculate (3.18). For holes in the valence band similar expressions can be written.

The conduction- and valence-band edges can now be drawn in Fig. 3.3. Our model contains p-type GaAs active layer whose doping concentration is $5 \cdot 10^{17}$ cm^{-3}. Using (3.16, 21), we now obtain

$$\delta_1 = F - E_v = -kT \ln(p/N_v) = 0.072 \text{ eV} \, , \tag{3.22}$$

$$\delta_2 = E_c - F = -kT \, [\ln(n/N_c) + 3.53553 \cdot 10^{-1} (n/N_c)$$
$$- 4.95009 \cdot 10^{-3} (n/N_c)^2] = 0.037 \text{ eV} \, . \tag{3.23}$$

(iv) Determination of V_D

The built-in potential V_D is deduced from (3.3) as

$$V_D = (1.424 + 0.318 - 0.037 - 0.072\text{eV})/e = 1.633 \text{ V} \, . \tag{3.24}$$

(v) Determination of the distance of the space-charge region.

The distances x_1 and x_2 from the p-n junction have been obtained and listed in Table 3.1.

Here, the dielectric constant of $Ga_{1-x}Al_xAs$ is

$$\epsilon(Ga_{1-x}Al_xAs) = (13.1 - 3.0x)\epsilon_0 \, , \quad \epsilon_0 = 8.85 \cdot 10^{-14} \text{ F/cm} \, . \tag{3.25}$$

Therefore, in our case,

$$\epsilon_1 = 1.16 \cdot 10^{-12} \text{ F/cm} , \quad \epsilon_2 = 3.63 \cdot 10^{-13} \text{ F/cm} . \tag{3.26}$$

The built-in potential V_D is disignated as V_{D_1} in region 1 (p-type GaAs); and V_{D_2} in region 2 (n-type $Ga_{0.7}Al_{0.3As}$), i.e.,

$$V_{D_1} = V_D/K , \quad V_{D_2} = V_D - V_{D_1} \tag{3.27}$$

where

$$K = 1 + \frac{\epsilon_1 (N_{A1}^- - N_{D1}^+)}{\epsilon_2 (N_{D2}^+ - N_{A2}^-)} , \tag{3.28}$$

N_A^- being the ionized acceptor concentration, and N_D^+ the ionized donor concentration.

The energy difference between the donor level and the conduction-band edge is so small, ≈ 0.006 eV, that all of the donors are ionized if n > $2 \cdot 10^{16}$ cm^{-3}; therefore, $(N_D^+ - N_A^-)$ of n-type GaAs in the space-charge region is equal to the carrier concentration n while $(N_A^- - N_D^+)$ of p-type GaAs can be assumed to be 1.1p [3.10]. $(N_D^+ - N_A^-)$ of n-type $Ga_{0.7}Al_{0.3}$ As can be approximately twice the carrier concentration n, therefore $(N_D^+ - N_A^-) = 2n$ [3.11]. However, $(N_A^- - N_D^+)$ of p-type GaAlAs requires a more complicated calculation. An example is the graphical solution employed by *Casey* and *Panish* [3.11]. Using (3.28), K is obtained by

$$K = 1 + \frac{1.16 \cdot 10^{-12} \times 1.1 \cdot 5 \cdot 10^{17}}{3.63 \cdot 10^{-13} \times 2 \cdot 2 \cdot 10^{17}} = 5.39 . \tag{3.29}$$

Then,

$$V_{D1} = 0.303 \text{ V} , \quad V_{D2} = 1.33 \text{ V} . \tag{3.30}$$

The applied voltage is distributed as in (3.27)

$$V_{a1} = V_a/K , \quad V_{a2} = V_a - V_{a1} . \tag{3.31}$$

The edge of the space-charge region in region 1 is

$$x_1 = \frac{[2\epsilon_1 (V_{D1} - V_{A1})]^{1/2}}{[e(N_A^- - N_D^+)]^{1/2}} . \tag{3.32}$$

For zero bias ($V_a = 0$) and

$$x_1 = \frac{(2 \times 13.1 \times 8.85 \cdot 10^{-14} \cdot 0.303)^{1/2}}{(1.6 \cdot 10^{-19} \cdot 1.1 \times 5 \cdot 10^{17})^{1/2}} = 0.028 \ \mu m \ , \tag{3.33}$$

$$x_2 = \frac{[2\epsilon_2 (V_{D2} - V_{a2})]^{1/2}}{[e(N_D^+ - N_A^-)]^{1/2}} = 0.039 \ \mu m \ . \tag{3.34}$$

Hence, the profile of the energy-band diagram is listed in Table 3.1

$$E_{v1} = V_{D1}[1 - (1 + x/x_1)^2] \qquad (-x_1 \leq x \leq 0) \ , \tag{3.35}$$
$$E_{c1} = V_{D1}[1 - (1 - x/x_1)^2] \qquad (-x_1 \leq x \leq 0) \ , \tag{3.36}$$
$$E_{v2} = -\Delta E_v - V_{D2}[1 - (1 - x/x_2)^2] \qquad (0 \leq x \leq x_2) \ , \tag{3.37}$$
$$E_{c2} = \Delta E_c - V_{D2}[1 - (1 - x/x_2)^2] \qquad (0 \leq x \leq x_2) \ . \tag{3.38}$$

The energy-band diagram of the p-n junction is shown in Fig. 3.4.

Let us move on to a p-p junction. Band-gap energies and hetero-barriers are given by

$$Eg_0 = 1.798 \ eV \ , \quad Eg_1 = 1.424 \ eV \ ,$$
$$\Delta E_c = 0.318 \ eV \ , \quad \Delta E_v = 0.056 \ eV \ . \tag{3.39}$$

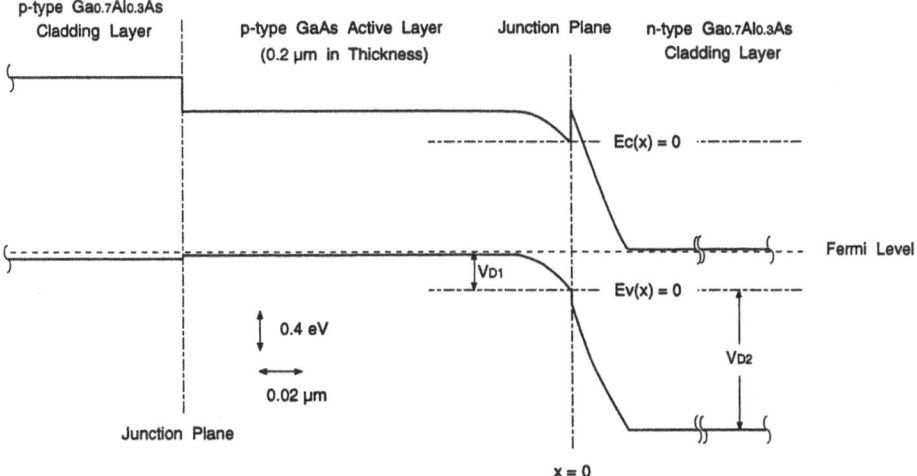

Fig. 3.4. Energy-band diagram of p-type-$Ga_{0.7}Al_{0.3}$As/p-type-GaAs/n-type-$Ga_{0.7}Al_{0.3}$As

From (3.15)

$$\delta_0 = F - E_v = -kT\ln(p/N_v) = 0.10 \text{ eV}, \qquad (3.40)$$

$$V_D = (F_0 - F_1)/e = (\Delta E_v - \delta_0 + \delta_1)/e = 0.028 \text{ V}, \qquad (3.41)$$

where δ_1 has been obtained in (3.22). Thus, the Fermi levels of both sides of the barrier are nearly coincident before they become connected. The built-in potential can then be neglected in this case. Figure 3.4 illustrates the detailed energy-band diagram.

3.1.4 Optical Properties of DH Lasers

a) Step-Index Planar Waveguide

The simplest optical model for DH lasers is the symmetric, three-layer slab waveguide depicted in Fig.3.5. The refractive indices of the core and cladding layers are designated as n_1 and n_2, respectively, and the thickness of the core is 2a.

We start out with deriving wave equations for the dielectric slab waveguide. Assuming a time and distance dependence of the form $\exp[j(\omega t - \beta z)]$, Maxwell's equations are

$$\nabla \times \mathbf{E} = -j\omega\mu\mathbf{H}, \qquad (3.42)$$

$$\nabla \times \mathbf{H} = j\omega\epsilon\mathbf{E}, \qquad (3.43)$$

$$\nabla \cdot \mathbf{D} = 0, \qquad (3.44)$$

$$\nabla \cdot \mathbf{B} = 0. \qquad (3.45)$$

With the help of a vector formula we have

$$\nabla \times \nabla \times \mathbf{E} = \nabla(\nabla \cdot \mathbf{E}) - \nabla^2 \mathbf{E} \qquad (3.46)$$

where ∇^2 denotes the Laplacian operator. Substituting (3.42 and 44) into (3.46), the wave equation for E can be expressed as

$$\nabla^2 \mathbf{E} + \omega^2 \epsilon\mu\mathbf{E} = -\nabla\left[\mathbf{E}\frac{\nabla\epsilon}{\epsilon}\right]. \qquad (3.47)$$

A similar wave equation is derived from (3.43 and 45) along the same lines, i.e.,

$$\nabla^2 \mathbf{H} + \omega^2 \epsilon\mu\mathbf{H} = -\frac{\nabla\epsilon}{\epsilon} \times (\nabla \times \mathbf{H}). \qquad (3.48)$$

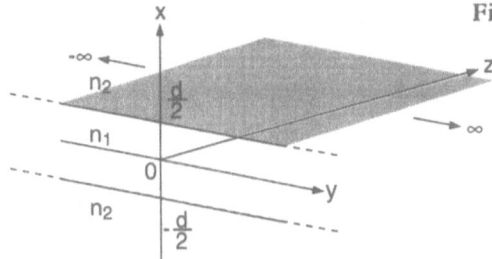

Fig.3.5. A step-index planar waveguide

In the present case, since $\nabla \epsilon = 0$, we obtain the next wave equation for the E_z and H_z components:

$$\frac{\partial^2 E_z}{\partial x^2} + (k_0^2 n^2 - \beta^2) E_z = 0 , \tag{3.49}$$

$$\frac{\partial^2 H_z}{\partial x^2} + (k_0^2 n^2 - \beta^2) H_z = 0 , \tag{3.50}$$

where $\mu = \mu_0$, $n^2 = \epsilon/\epsilon_0$ and $k_0^2 = \omega^2 \epsilon_0 \mu_0$.

The transverse-field components are expressed in terms of E_z and H_z with the aid of (3.42 and 43) as follows:

$$E_x = -\frac{j}{\omega^2 \epsilon\mu - \beta^2} \left[\beta \frac{\partial E_z}{\partial x} + \omega\mu \frac{\partial H_z}{\partial y} \right] , \tag{3.51}$$

$$E_y = \frac{j}{\omega^2 \epsilon\mu - \beta^2} \left[-\beta \frac{\partial E_z}{\partial y} + \omega\mu \frac{\partial H_z}{\partial x} \right] , \tag{3.52}$$

$$H_x = \frac{j}{\omega^2 \epsilon\mu - \beta^2} \left[\omega\epsilon \frac{\partial E_z}{\partial y} - \beta \frac{\partial H_z}{\partial x} \right] , \tag{3.53}$$

$$H_y = -\frac{j}{\omega^2 \epsilon\mu - \beta^2} \left[\omega\epsilon \frac{\partial E_z}{\partial x} + \beta \frac{\partial H_z}{\partial y} \right] . \tag{3.54}$$

The guided modes must satisfy (3.49 and 50). However, because these equations are independent of each other, general solutions of (3.49 and 50) can be expressed by a sum of two independent solutions; that is, one satisfies (3.49) with $H_z = 0$ and (3.50) with $E_z = 0$. The latter solutions are the *TE modes*, because only the transverse-electric field E_y is different zero. In the same way, the former solutions are called *TM modes*. The waveguide is considered infinite in extent in the y direction so that $\partial/\partial y = 0$. Then, for $\partial/\partial y = 0$, the following field components remain:

TE modes $\quad$ E$(0,E_y,0)$, $\quad$ H$(H_x,0,H_z)$,

TM modes $\quad$ E$(E_x,0,E_z)$, $\quad$ H$(0,H_y,0)$.

b) TE Modes

The tangential field compoents of the TE modes are E_y and H_z. Since these two are related directly to the boundary conditions, it is convenient to deal with the E_y component instead of E_z. From (3.48), E_y must satisfy

$$\frac{\partial^2 E_y}{\partial x^2} + (k_0^2 n_1^2 - \beta^2)E_y = 0 \quad \text{(core)}, \tag{3.55}$$

$$\frac{\partial^2 E_y}{\partial x^2} + (k_0^2 n_2^2 - \beta^2)E_y = 0 \quad \text{(cladding)}. \tag{3.56}$$

In the core the special solutions of (3.55) are the cosine and sine functions.

In the cladding layers of a DH laser the solutions are classified into two types: the evanescent (exponentially decaying) solution for $n_2 k < \beta < n_1 k$, called the *guided mode*; and the sinusoidal oscillating solution for $\beta < n_2 k$. A portion of the optical power of a guided mode is essentially confined to the core[1], while the remaining power emerges from the cladding layer. As the power radiates from the cladding layer, it is no longer confined to the core and produces a set of radiation modes. The group of all of the guided and radiation modes constitutes a complete orthogonal set and any field can be expanded in terms of these guided and radiation modes.

The solution of a guided mode must satisfy the boundary conditions stating that the tangential components of the electric field must be continuous at the core-cladding boundary and approach zero as $x \rightarrow \infty$. From the above conditions, we obtain the following mode distributions

$(Even)\quad E_y(x) = A_e \cos(\kappa x) \quad |x| \leq d/2$
$$\qquad\qquad = A_e \cos(\kappa d/2)\exp[-\gamma(|x|-d/2)] \quad |x| \geq d/2 \tag{3.57}$$

$(Odd)\quad E_y(x) = A_0 \sin(\kappa x) \quad |x| \leq d/2$
$$\qquad\qquad = (x/|x|) A_0 \sin(\kappa d/2)\exp[-\gamma(|x|-d/2)] \quad |x| \geq d/2 \tag{3.58}$$

where

$$\kappa^2 = k_0^2 n_1 - \beta^2, \quad \gamma^2 = \beta^2 - k_0^2 n_2^2. \tag{3.59,60}$$

---

[1] The effective core thickness is somewhat larger than the geometrical core size due to the Goos-Hänchen effect [3.12].

Table 3.2. Mode functions of step-index planar waveguides (d = 2a)

	Even mode	Odd mode	Deterministic equation
TE	$E_y = \begin{cases} A_e \cos\kappa x, & (|x| \leq a) \\ A_e \cos(\kappa a)e^{-\gamma(|x|-a)}, & (|x| \geq a) \end{cases}$	$E_y = \begin{cases} A_0 \sin\kappa x, & (|x| \leq a) \\ \dfrac{x}{|x|} A_0 \sin(\kappa a)e^{-\gamma(|x|-a)}, & (|x| \geq a) \end{cases}$	$(\kappa a)^2 + (\gamma a)^2 = V^2$ $\begin{cases} \tan(\kappa a) = \dfrac{\gamma a}{\kappa a} \text{ (even order)} \\[2mm] \tan(\kappa a) = -\dfrac{\kappa a}{\gamma a} \text{ (odd order)} \end{cases}$
	$d = 2a$		
TM	$H_y = \begin{cases} B_e \cos\kappa x, & (|x| \leq a) \\ B_e \cos(\kappa a)e^{-\gamma(|x|-a)}, & (|x| \geq a) \end{cases}$	$H_y = \begin{cases} B_0 \sin\kappa x, & (|x| \leq a) \\ \dfrac{x}{|x|} B_0 \sin(\kappa a)e^{-\gamma(|x|-a)}, & (|x| \geq a) \end{cases}$	$(\kappa a)^2 + (\gamma a)^2 = V^2$ $\begin{cases} \tan(\kappa a) = \left(\dfrac{n_1}{n_2}\right)^2 \dfrac{\gamma a}{\kappa a} \text{ (even order)} \\[2mm] \tan(\kappa a) = -\left(\dfrac{n_2}{n_1}\right)^2 \dfrac{\kappa a}{\gamma a} \text{ (odd order)} \end{cases}$

Under these conditions H_z is another tangential field component continuous at the core-cladding boundary. From this boundary condition, the eigenvalue equations for the even and odd TE modes are derived as in (3.61, 62), respectively; they are tabulated in Table 3.2, i.e.,

$$\tan(\kappa d/2) = \frac{\gamma d/2}{\kappa d/2} \quad \text{(even)} \tag{3.61}$$

$$-\tan(\kappa d/2) = \frac{\kappa d/2}{\gamma d/2} \quad \text{(odd)} \tag{3.62}$$

The solutions of these eigenvalue equations can be normalized by introducing the new parameters b and V defined by

$$b = \frac{\gamma^2}{\kappa^2 + \gamma^2} = \frac{(\beta/k_0)^2 - (n_2)^2}{n_1^2 - n_2^2} \tag{3.63}$$

$$V = \sqrt{(\kappa d/2)^2 + (\gamma d/2)^2} = k_0(d/2)n_1\sqrt{2\Delta} \tag{3.64}$$

where

$$\Delta = \frac{n_1^2 - n_2^2}{2n_1^2} \simeq \frac{n_1 - n_2}{n_1}, \tag{3.65}$$

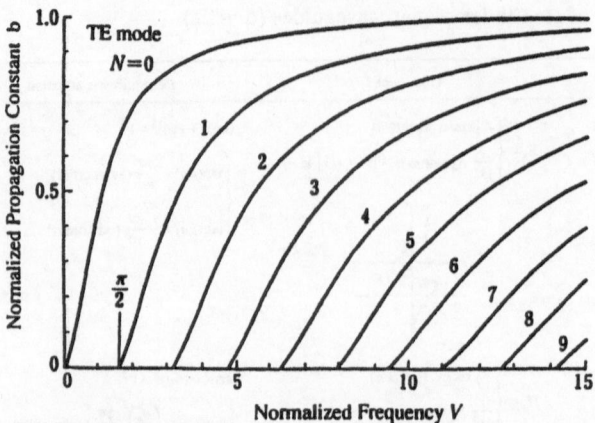

Fig.3.6. Dispersion curves of some TE modes

and yield the simple closed form

$$V = \frac{\pi/2}{\sqrt{1-b}}\left[(2/\pi)\tan^{-1}\sqrt{\frac{b}{1-b}} + N\right], \tag{3.66}$$

$$\kappa(d/2) = V\sqrt{1-b}, \tag{3.67}$$

$$\gamma(d/2) = V\sqrt{b}. \tag{3.68}$$

The dispersion curve relates V and b, as shown in Fig.3.6. When the waveguide parameters n_1, n_2, d, and the wavelength λ are given, the propagation constant β_N of any mode can be obtained from (3.63-68). The mode number is labeled in the order of increasing N, and TE_0, where $N = 0$, designates the *fundamental mode*. The mode number N corresponds to the number of nodes in the field distribution. The electric field of the model (d $= 0.2\,\mu$m) in Fig.3.5 can be drawn as in Fig.3.7a. Figure 3.7b exhibits the electric fields of all modes which can exist when d $= 1.0\,\mu$m. When the propagation constant of one guided mode reaches $n_2 k(b\rightarrow 0)$, the mode is at "cutoff", and the V value is then called the *cutoff V-value*. Setting $\gamma = 0$ and $\kappa(d/2) = V$, the cutoff V-value of the TE modes is easily obtained from (3.68) as

$$V = (\pi/2)N \quad (N = 0, 1, 2, ...). \tag{3.69}$$

The cutoff V-value for the TE_1 gives the single-mode condition and, when V is smaller than $\pi/2$, only the TE_0 mode can propagate. The single-mode condition is important for designing single-mode waveguides and single-mode optical fibers with an arbitrary refractive-index profile.

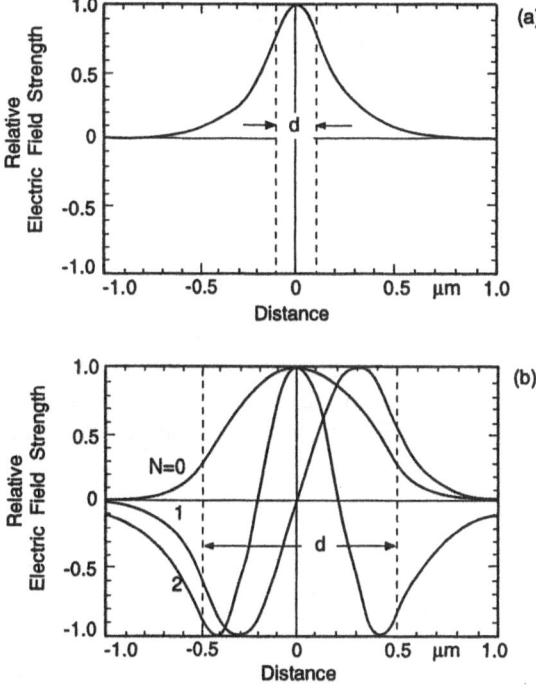

Fig.3.7a,b. The parameters are $n_1 = 3.590$, and $n_2 = 3.385$. **(a)** Fundamental TE mode for $d = 0.2$ μm. **(b)** Fundamental, first- and second-order TE mode for $d = 1.0$ μm [3.13]

c) TM Modes

Stating from $H_z = 0$, the H_y component of TM modes is obtained in the same way:

$(Even)$ $H_y(x) = B_e \cos\kappa x$ $|x| \leq d/2$

$\qquad\qquad = B_e \cos(\kappa d/2) \exp[-\gamma(|x| - d/2)]$ $|x| \geq d/2$,   (3.70)

(Odd) $H_y(x) = B_0 \sin(\kappa x)$ $|x| \leq d/2$

$\qquad\qquad = (x/|x|) B_0 \sin(\kappa d/2) \exp[-\gamma(|x| - d/2)]$ $|x| \geq d/2$.   (3.71)

Another tangential field component E_z is obtained by

$$E_z = \frac{1}{j\omega\epsilon} \frac{\partial H_y}{\partial x} .$$ (3.72)

Since E_z must be continuous at the core-cladding boundary, the eigen-value equations for even and odd TM modes are derived with the aid of (3.70 or 71) via (3.72) as

$$\tan(\kappa d/2) = \left(\frac{n_1}{n_2}\right)^2 \frac{\gamma d/2}{\kappa d/2} \quad \text{(even)}, \tag{3.73}$$

$$-\tan(\kappa d/2) = \left(\frac{n_2}{n_1}\right)^2 \frac{\kappa d/2}{\gamma d/2} \quad \text{(odd)}. \tag{3.74}$$

The results are listed in Table 3.2. The solutions of (3.73, 74) are normalized with the parameters b and V, given by (3.63, 64). The solutions for the TM modes differ from those for the TE modes due to the $(n_1/n_2)^2$ and $(n_2/n_1)^2$ terms in (3.61, 62).

$$V = \frac{\pi/2}{\sqrt{1-b}} \left[(2/\pi)(n_1/n_2)^2 \tan^{-1} \sqrt{\frac{b}{1-b}} + N \right]. \tag{3.75}$$

However, the difference between the solutions for the TM and TE modes is usually negligible because n_1/n_2 can be approximated to be unity in most cases. The cutoff V-values of the TM modes are therefore equal to those of the TE modes when the mode numbers are equal.

d) Mode-Confinement Factor

The mode-confinement factor ξ is defined as the ratio of the optical power confined in the core region to the total power. This factor is important for semiconductor lasers having optical gain in the core region, because it is related to the mode gain. The mode-confinement factor of an even TE mode is calculated from (3.57) to be

$$\xi = \frac{\int_0^{d/2} |E_y|^2 \, dx}{\int_0^\infty |E_y|^2 \, dx} = \frac{1 + [\sin(\kappa d)]/\kappa d}{1 + [\sin(\kappa d)]/\kappa d + 2[\cos^2(\kappa d/2)]/\gamma d}. \tag{3.76}$$

Applying (3.76) to our model ($d = 0.2\,\mu m$), the confinement factor ξ is 0.587, which means that 58.7% of the total power is confined to the core.

Since (3.76) is not normalized by the index difference and the waveguide thickness, it is not convenient for practical use. By employing (3.46, 63 and 64) we can simply express the confinement factor in terms of V and b as

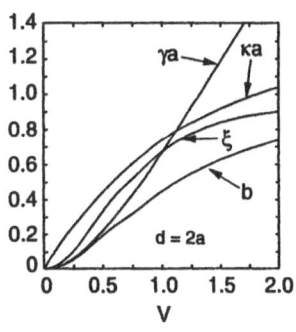

Fig.3.8. Confinement factors ξ and other parameters b, κa, and γa versus V

$$\xi = \frac{V + \sqrt{b}}{V + 1/\sqrt{b}} \tag{3.77}$$

The relationship between V and b is exhibited in Fig.3.6, so that the confinement factor can be expressed as a function of V. In Fig.3.7 we show a typical field distribution of some lower-order modes. The confinement factor ξ versus V is shown in Fig.3.8 with the parameters $\kappa d/2$, $\gamma d/2$, and b.

For convenience, the confinement factor of the TE_0 mode versus the active layer thickness of GaInAsP/InP lasers is depicted in Fig.3.9. In the case of GaInAsP/InP semiconductor lasers, the refractive index of the InP cladding layer is constant. On the other hand, the lasing wavelength and the refractive index of a GaInAsP active layer varies according to the compositions x and y. The wavelength and refractive-index difference therefore changes simultaneously in Fig.3.9.

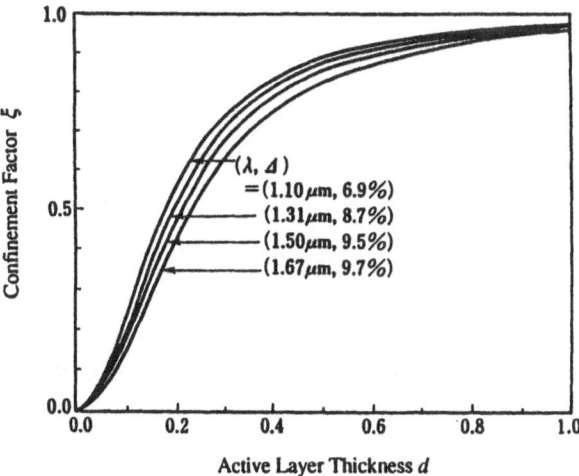

Fig.3.9. The mode confinement factor ξ [3.14]

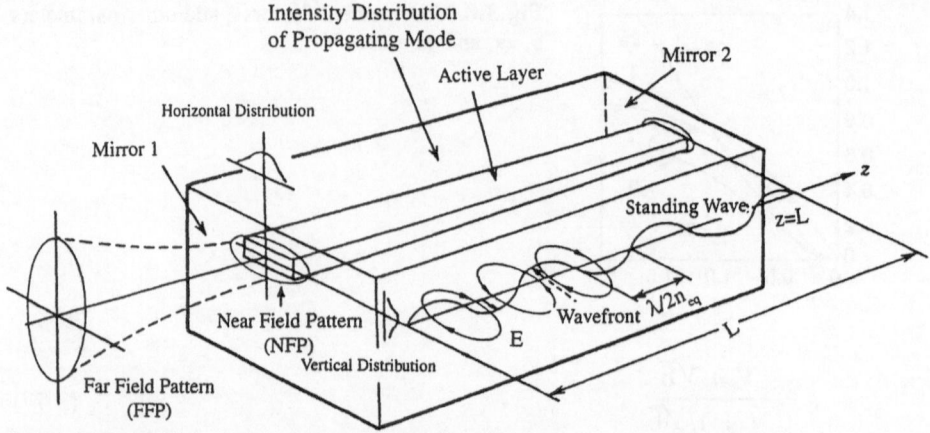

Fig.3.10. Intensity distribution and beam divergence of a semiconductor laser

3.1.5 Threshold Current of DH Lasers

Under laser-oscillation conditions light makes a round trip in the laser resonator, and a standing wave exists corresponding to a transverse mode with a wavefront essentially parallel to a reflecting mirror. A portion of the power leaves the resonator via transmission through the reflecting mirror.

Figure 3.10 displays a standing wave. Let us assume that the mode of interest experiences a power loss due to absorption in the semiconductor medium per unit length of α_{ac} [cm^{-1}] in the active layer and α_{cl} [cm^{-1}] in the cladding layer, and the amplitude reflectivities of the mirrors and the phase shifts are R_1, R_2, ϕ_1, and ϕ_2, respectively. When the mode is amplified by induced emission, the ratio is expressed by the power gain g [cm^{-1}]. Light begins propagating from z = 0, is reflected at z = L, and then returns again to z = 0. Since the electric field should not change, the resonance condition is written as

$$\exp\left[\xi g L - \xi\alpha_{ac}L - (1-\xi)\alpha_{cl}L - j2\beta L - j\phi_1 - j\phi_2\right]R_1 R_2 = 1 \qquad (3.78)$$

where β denotes the propagation constant, and ξg is the mode gain. Here, the gain which satisfies the resonance condition is called the *threshold gain* g_{th}. It is given by comparing the real parts of (3.63), i.e.,

$$g_{th} = \alpha_{ac} + \left[\frac{1}{\xi} - 1\right]\alpha_{cl} + \frac{1}{2\xi L}\ln\left[\frac{1}{R_1 R_2}\right]. \qquad (3.79)$$

The first and second terms on the right-hand side specify the absorption of the laser medium which is due to the absorption of free carriers, and has a value of about 10 cm^{-1} in GaAs. The third term arises because the reflectivity of the mirrors is less than 1.

The Fresnel reflection between a dielectric surface and air is written with the refractive index n as

$$R = \frac{(n-1)^2}{(n+1)^2} .$$
(3.80)

This equation has, strictly speaking, been derived for a plane wave. If the refractive index varies between the core and cladding layers, or if the propagation behavior of guided modes differs between the layers, the reflection coefficients between the TE and TM modes are dissimilar [3.15]. In this case the equivalent refractive index n_{eq} instead of n may be used. *Ikegami* [3.15] reported that the mirror reflectivity at the cleaved facet for the TE modes is higher than that for the TM modes. For this reason TE modes become more dominant as the threshold is approached, and the laser starts oscillating in the TE modes.

The imaginary part gives the resonance condition

$$2\beta L + \phi_1 + \phi_2 = 2\pi q \quad (q \text{ being an integer}) .$$
(3.81)

Equation (3.81) also applies to the case where reflecting mirrors are coated at the edges or diffraction gratings are added. If $\phi_1(\lambda)$, $\phi_2(\lambda)$ and $2\pi n_{eq}/\lambda$ instead of β are substituted into (3.66), we obtain

$$4\pi n_{eq} L/\lambda + \phi_1(\lambda) + \phi_2(\lambda) = 2\pi q .$$
(3.82)

As the injection current is increased, a specific longitudinal mode that satisfies the resonance condition starts oscillating and the output power increases. The point where the output power increases rapidly is called the *threshold*, and the corresponding current is termed the *threshold current*. The threshold current divided by the area of the active region is called the *threshold current density*. If we assume a linear gain versus an injected carrier density n, then we obtain

$$g = A_0 n - \alpha_{in}$$
(3.83)

where A_0 is the linear gain constant, and α_{in} is the residual loss. The threshold current density is described [3.16] by

$$J = edn/\tau_s \tag{3.84}$$

with

$$\tau_s = 1/(B_{eff}n) \tag{3.85}$$

where τ_s and B_{eff} are the carrier lifetime and the effective recombination constant, respectively. From (3.79, 83-85) the threshold current density J_{th} is

$$J_{th} = \frac{edB_{eff}}{A_0{}^2}\left[\alpha_{in} + \alpha_{ac} + \left(\frac{1}{\xi} - 1\right)\alpha_{cl} + \frac{1}{2\xi L}\ln\left(\frac{1}{R_1 R_2}\right)\right]^2 . \tag{3.86}$$

The normalized threshold-current density J_{th}/d is the threshold-current density divided by the active layer thickness. It is a very important factor which characterizes the quality of a DH structure. A detailed comparison between the theoretically described optical properties of DH lasers and the experimental results is given in Chap. 9.

4. Epitaxy of III-V Compound Semiconductors

This chapter summarizes the fundamental concepts of the epitaxial growth of III-V compound semiconductors. The relevant crystal-growth technologies encompass two categories: bulk crystal preparation for a substrate to be used as a seed, and thin-film epitaxial growth. Later chapters will contain detailed discussion of the epitaxial-growth techniques, and here we overview several characteristics of liquid-phase epitaxy by comparing them with other crystal-growth techniques, such as Metalorganic Chemical Vapor Deposition (MOCVD), Molecular Beam Epitaxy (MBE), and Chemical Beam Epitaxy (CBE), among others.

4.1 III-V Substrates for Semiconductor Lasers

4.1.1 Necessity of Substrates

A substrate can be thought of as a seed crystal on which a double heterostructure is formed. Therefore, the quality of the substrate used may largely influence the device characteristics, especially in the case of active devices or an Opto-Electronic Integrated Circuit (OEIC). Furthermore, since the total thickness of epitaxial layers is normally about 5 μm, the substrate serves as an important reinforcement of the laser structure. We look at the substrate requirements first.

4.1.2 Substrate Quality Requirements

The important characteristics of the substrate are as follows:
- Crystal direction.
- Etch pit density.
- Impurity concentration.
- Substrate thickness.
- Wafer size.

GaAs and InP are popularly used as substrates for optical devices. Generally speaking, binary materials are employed for substrates due to the availability of good crystals fulfilling the aforementioned requirements. The

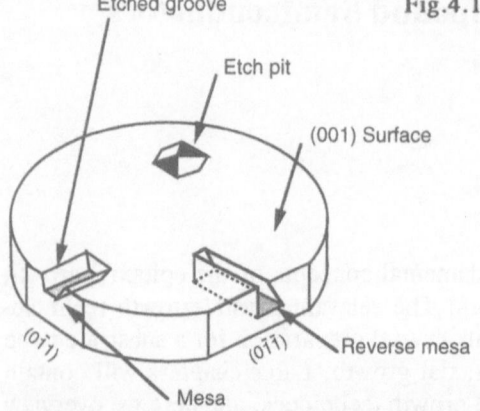

Fig.4.1. Crystal orientation of substrates [4.2]

Etched groove

Etch pit

(001) Surface

(0̄11)

(01̄1̄)

Reverse mesa

Mesa

(100) oriented substrates are commonly displayed, since (011) or (0̄11) cleaved facets are used as laser mirrors, as shown in Fig.4.1. A (100) facet deviation within $\pm 0.1°$ or $\pm 0.5°$ is now available. On the other hand, 2°-off substrates have come to be popular for MOCVD crystal growth. Residual imperfections can be checked by selectively etching the substrate. One of the crystal imperfections that may appear on an etched (100) surface are rectangular hills or holes, the so-called *etch pits*, also shown in Fig.4.1. A solution of $HF:CrO_3 + H_2O = 2:1$ is used as a selective etch for GaAs. CrO_3 is dissolved in water at the rate of 33 wt.%. The Etch Pit Density (EPD) represents the number of etch pits per square centimeter. The spatial distribution of EPD on the substrate varies according to the crystal-growth technique. GaAs substrates with an EPD less than 2 000 are sometimes used for commercial laser devices. Bulk growth techniques for InP have been improved and now EPD of less than 50 000 are common. A solution of $HCl: H_2O = 4:1$ is used on InP to reveal the imperfections.

The popular dopants for n-type and p-type GaAs substrates are Si and Zn, respectively. We can discuss dopants by referring to the Periodic Table (Table 4.1). If a group-II material replaces a group-III material, for instance Zn replaces Ga, then Zn requires an additional electron to bond with Ga atoms. In this case, Zn acts as an acceptor. On the other hand, a group-VI

Table 4.1. Periodic Table [4.1]

IA	IIA	IIB	IIIA	IVA	VA	VIA	VIIA	VIIIA
H								He
Li	Be		B	C	N	O	F	Ne
Na	Mg		Al	Si	P	S	Cl	Ar
K	Ca	Zn	Ga	Ge	As	Se	Br	Kr
Rb	Sr	Cd	In	Sn	Sb	Te	I	Xe
Cs	Ba	Hg	Tl	Pb	Bi	Po	At	Rn
Fr	Ra							

material such as Te can act as a donor when it replaces a group-V material such as As due to the excess Te electron. Note that a group-IV material such as Si can occupy either a Ga site or an As site, depending on the crystal-growth conditions. Si becomes a donor when it replaces Ga, and it becomes an acceptor when it replaces As. S and Sn are commonly used for n-type dopants of InP substrates, and Zn is a p-type dopant. The impurity concentrations for laser diodes are usually designed to be around 10^{18} cm^{-3}.

A semi-insulating substrate, whose common definition is that its resistivity exceeds 10^6 $\Omega \cdot cm$, is very important, especially for OEICs.

Substrates 5 or 7.5 cm (i.e., 2 or 3 inches) in diameter are now commercially available. Each substrate is sliced from a rod and is approximately 350 μm thick. The surface of the substrate is mechanochemically polished. Some crystal damage might remain, and the surface must be etched before epitaxial growth in order to eliminate the damaged layer.

4.2 Bulk Growth Techniques

Bulk crystals are grown by cooling and solidifying the solution associated with the III-V compound semiconductors to be grown. Each element of III-V compound semiconductors has a different vapor pressure, e.g., the vapor pressure of As is over 4 orders of magnitude higher than that of Ga at 1238°C, the melting point of GaAs. Therefore, a stoichiometric solid composition, Ga:As = 1:1, is impossible, if Ga and As are equally mixed at atmospheric pressure above the melting temperature. Techniques must be introduced during bulk crystal growth to increase the presence of elements with a higher vapor pressure.

The Horizontal Bridgeman (HB) or the Czochralski methods are popular bulk crystal-growth techniques. The HB method is used to produce multicrystal source materials with higher purity. The LEC (Liquid EnCapsulate) method is one of the Czochralski methods and is utilized for single-crystal wafers. Figure 4.2 outlines the LEC method.

4.3 Heteroepitaxial Techniques

4.3.1 Liquid-Phase Epitaxy

The principle of Liquid-Phase Epitaxy (LPE) is very simple. Epitaxial layers which have the same or a certain fixed crystal-growth direction against that of the substrate can be thermodynamically grown on the substrate when the substrate comes into contact with the supersaturated solution. Figure 4.3

Fig.4.2. Outline of LEC method

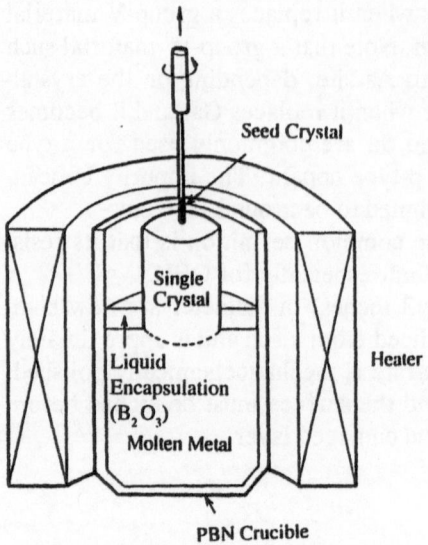

Fig.4.2. Outline of LEC method

schematically represents the liquidus-solidus equilibrium in a system with a large AC primary-phase field. For instance, if we assume A and B to be Ga and As, then T_A, T_B and T_{AB} represent the melting points of Ga, As and GaAs, which are 29.8°C, 810°C and 1238°C, respectively. A Ga-rich Ga-As melt of which the As atom fraction is X_2 saturates at T_2. A supersaturated GaAs in the solution is separated as a crystal on a GaAs substrate as the temperature decreases from T_2 to T_1. The LPE technique is very popular because of its high productivity of commercial devices, especially for the epitaxial growth of semiconductor lasers. It has been used since the advent of semiconductor-laser research.

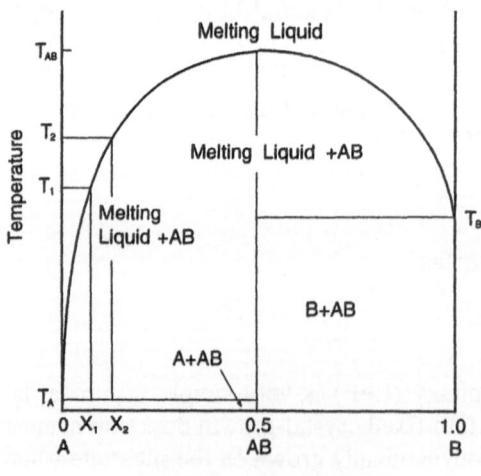

Fig.4.3. Correlative figure of AB chemical compound

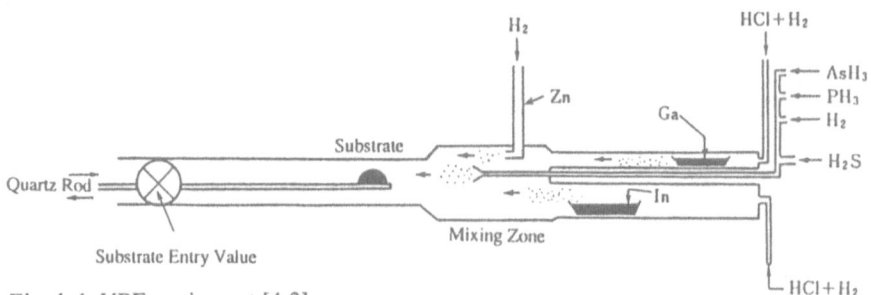

Fig.4.4. VPE equipment [4.3]

4.3.2 Vapor-Phase Epitaxy

A Vapor Phase-Epitaxy (VPE) or Chemical-Vapor Deposition (CVD) technique enables crystal growth from materials in the vapor state. Figure 4.4 shows an example of VPE equipment which was reported by *Neuse* (RCA) [4.3]. A lifetime in excess of several thousand hours in a $Ga_x In_{1-x} As_y \cdot P_{1-y}/InP$ laser, and the crystal growth of the longest wavelength ($y = 1$) were reported by *Neuse* [4.3]. NEC and Bell Laboratories developed a new method to produce a sharp transition between layers by dividing a reaction tube into two parts [4.4]. This VPE method permits the formation of wafers for GaInAsP/InP lasers with wavelengths of $1.1 \div 1.55$ μm and a threshold current density of approximately 1 kA/cm^2 in the active layer, with a thickness of d $= 0.1 \div 0.2$ μm. VPE is worthy of attention because of its capacity to produce large-sized wafers as well as controlling the film thickness for the application of optical integrated circuits.

4.3.3 Metalo-Organic Chemical-Vapor Deposition

Metalo-Organic Chemical-Vapor Deposition (MOCVD) is one of the most advanced growth techniques of vapor phase epitaxy, and is used to grow electronic devices which utilize quantum effects by superlattice or quantum-well structures, as well as homogeneously growing layers on large substrates. The system is distinctive in its use of metal-organic materials which have a fairly high vapor pressure, for example, TMG, TEG, TMA, TEA and TEI[1]. For the epitaxial growth of a GaAlAs/GaAs system, TMG or TEG, TEA and AsH$_3$ are utilized as gas sources, and are transported into the reactor through stainless steel tubes. These gases are heated to around $500 \div 800$°C to be decomposed and mixed very near to the substrate and are crystallized on the substrate. The chemical reactions are [4.5]

---

[1] TMG: TriMethylGallium, TEG: TriEthylGallium, TMA: TriMethylAluminum, TEA: Tri-EthylAluminum, TEI: TriEthylIndium

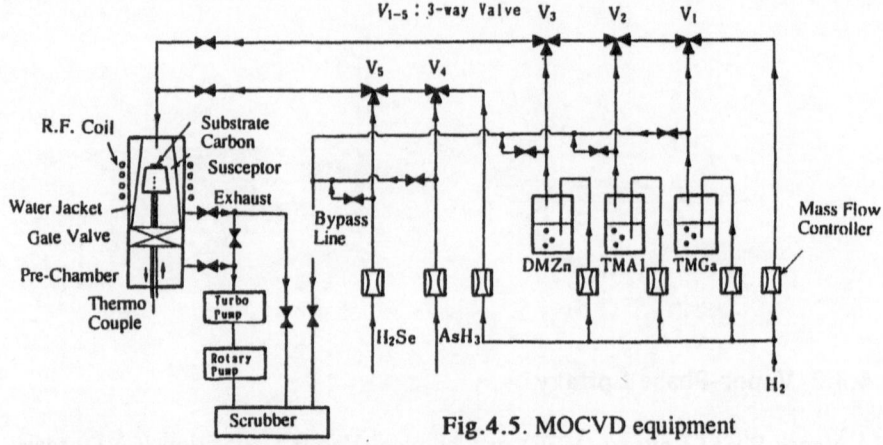

Fig.4.5. MOCVD equipment

$$(CH_3)_3 Ga + AsH_3 \rightarrow GaAs + 3CH_4 , \qquad (4.1)$$

$$(CH_3)_3 Al + AsH_3 \rightarrow AlAs + 3CH_4 . \qquad (4.2)$$

An extremely thin epitaxial layer such as a $1 \div 10$ nm superlattice or quantum-well structure can be grown with less than a 0.5 nm abrupt transition layer because of the very high-speed gas flow used in MOCVD. Figure 4.5 depicts an example of MOCVD equipment. In GaInAsP crystals, Thomson-CSF [4.6] reported a pulsed oscillation of a quaternary laser of 1.15 μm (J_{th} = 5.9kA/cm^2). Following this report, crystals of good quality are currently available for the lasing region from 1.3 to 1.6 μm.

As for GaAlAs/GaAs crystals and crystals in the visible region, a laser was initially realized on an experimental basis by *Dupuis* of Rockwell [4.7] concurrently with the Thomson-CSF report. CW operation with a lifetime of 1800 hours was reported in 1979 and the formation of devices with high reliability was suggested. A continuous test of 500 hours was carried out by STL in 1979, and the record was revised to 8000 hours the next year. Lasers with excellent performance are also fabricated by Thomson-CSF, SONY and Toshiba. Since 1985, efforts have been made using MOCVD to grow a semiconductor laser with a shorter wavelength ($\lambda = 0.63 \div 0.67 \mu$m), such as InGaAlP.

4.3.4 Molecular Beam Epitaxy

Molecular Beam Epitaxy (MBE) is an ultralow-pressure process ($\approx 10^{-10}$ Torr) in which beam-like molecules of several kinds of materials leave the heated effusion cells and impinge on the heated substrate, as depicted in Fig.4.6. Crystallization can be controlled by the shutters in front of the effusion cells, allowing an extremely abrupt transition layer as narrow as sev-

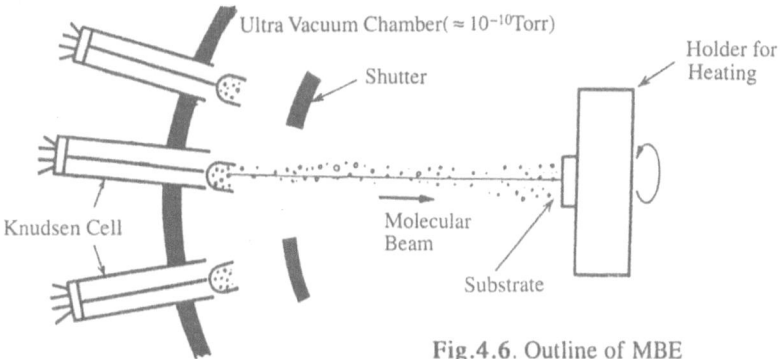

Fig.4.6. Outline of MBE

eral Ångstroms. The new function device which uses superlattice or quantum-well structures are expected to be realized by the MBE technology. But a MBE system is more expensive than the other crystal-growth systems because it needs an ultrahigh-vacuum chamber. *Tsang* of AT&T Bell Laboratories [4.8] produced GaAs wafers with the low threshold-current density of 800 A/cm^2, creating a practical laser with 10 000 hours, which is said to be superior to the laser by LPE in its threshold current, efficiency and temperature characteristics.

As for GaInAs, *Miller* et al. of AT&T Bell Laboratories [4.9] reported a room-temperature pulsed oscillation at $\lambda = 1.67 \ \mu$m in 1978. The value $J_{th}/d = 5$ kA/(cm$^2 \cdot \mu$m), was obtained by using a diffused p-n junction. Subsequently, oscillation of 1.7 μm lasers with a p-type cladding layer was attempted. GaInAs is the ternary material which corresponds to the longest wavelength of $Ga_x In_{1-x} As_y P_{1-y}$, i.e. 1.67 μm. Crystal growth of InP on GaInAs using LPE is difficult due to the occurrence of meltback. Due to the large difference of vapor pressures of As and P, it is very difficult to grow $Ga_x In_{1-x} As_y P_{1-y}$ using MBE. So, for GaInAsP a new epitaxial method had to be developed later.

4.3.5 Chemical Beam Epitaxy

To overcome the difficulty of MBE for a quaternary growth including P, Chemical Beam Epitaxy (CBE) was conceived by *Tsang* of AT&T Bell Laboratories [4.10]. Figure 4.7 illustrates an example of CBE equipment, which is similar to the MBE equipment except that gaseous materials are used entirely. Molecules of materials which are resolved at a high temperature reach a substrate and begin to grow. As a compensation method for the disadvantages of MBE, CBE is effective especially for a GaInAsP system. CBE is also known as gas-source MBE or metal-organic MBE [4.11].

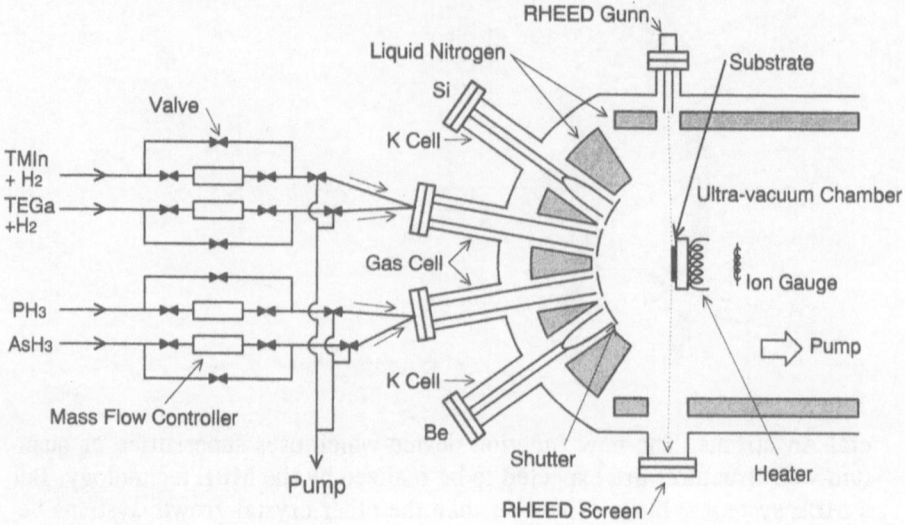

Fig.4.7. CBE equipment [4.10]

5. Liquid Phase Epitaxy and Growth Technology

Liquid-Phase Epitaxy (LPE) has many advantages and is capable of producing reliable semiconductor devices. This chapter provides a detailed description of the equipment necessary for LPE and the epitaxial techniques required for double heterostructures. As discussed in Chap. 4, the principle behind LPE is very simple: epitaxial layers can be thermodynamically grown on a substrate with the same orientation when the substrate meets the oversaturated solution. Since semiconductor-laser research began, LPE has been a popular growth technique and it still plays an important role in producing semiconductor lasers, light emitting diodes and photodetectors.

5.1 Outline of an LPE System

The LPE system consists of a reactor, a substrate loading subsystem, a pump and exhaust sub-system, a gas flow sub-system and a temperature control sub-system, as illustrated in Fig. 5.1. The reactor in which epitaxial layers are grown is the main part of the LPE system. Pure materials for making the reactor should be carefully chosen so as not to pollute the epitaxial layers with impurities during the crystal growth. The function of the substrate loading sub-system is to introduce and remove the substrate and metal solutions into and out of the reactor. The loading box is usually filled with nitrogen gas to purge the air while opening the reactor. The pump and exhaust sub-system pump air or impure gases from the reactor immediately

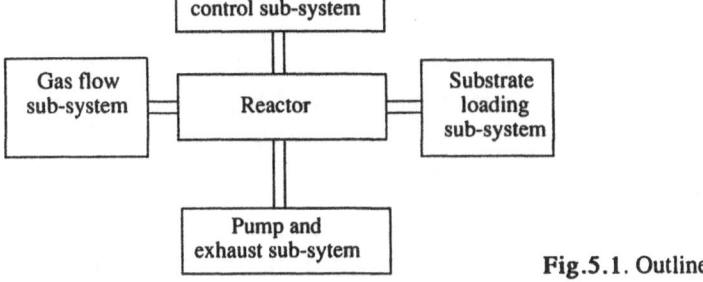

Fig. 5.1. Outline of LPE system.

after the materials are placed inside or taken out. For safety reasons, an exhaust line should be guided to the outdoors to eliminate the hydrogen gas exhaust from the room. The gas flow sub-system allows purified hydrogen gas to flow through the reactor while it is being heated in order to remove the mixture of impure gases. Hydrogen works especially well in promptly eliminating oxygen by chemical reaction at high temperatures. The temperature control sub-system heats up and cools down the reactor according to a given temperature program. An important factor of the LPE system is maintaining a uniform temperature distribution along the graphite boat precisely decreasing the reactor temperature at a certain rate while keeping the temperature constant. LPE systems should be designed to be operated by one person and to be easily repaired.

5.2 Reactors

Figure 5.2a,b illustrates the two types of reactors that are usually employed, i.e., horizontal and vertical systems. A horizontal reactor has widely been utilized in production lines and research laboratories. A three-zone horizontal heater fulfilling the requirements mentioned in Sect.5.1 has been developed. The typical temperature variation is within $\pm 0.1°C$ over an axial distance of 20 cm or more. The three-zone heater is made possible by independently adjusting the temperature distribution of each zone.

In contrast, the vertical reactor consists of a vertically placed quartz tube containing a rotary graphite boat, and it requires less space than the horizontal reactor. Temperature uniformity around the axis of the heater is necessary because the semiconductor materials are concentrically placed in the reactor. A slight vertical temperature gradient is introduced inside the heater in order to maintain the temperature uniformity in the vertical direction.

5.2.1 Horizontal Reactor

We illustrate the appearance of the horizontal reactor by referring to the authors' experimental apparatus. A quartz tube of 55 mm diameter, 170 cm length and 1 mm thickness is horizontally mounted. Both ends of the tube are polished to meet a vacuum joint made of rubber gaskets. The rubber gasket is pressed together by two flanges and fills up the opening between the tube and flanges without damaging the fragile tube. The rubber gasket, therefore, is often used to join a quartz tube or a vacuum gauge to a metal flange. Be aware that the rubber gasket is weak at high temperatures. Hence, a water jacket is necessary to keep the rubber gasket cool when

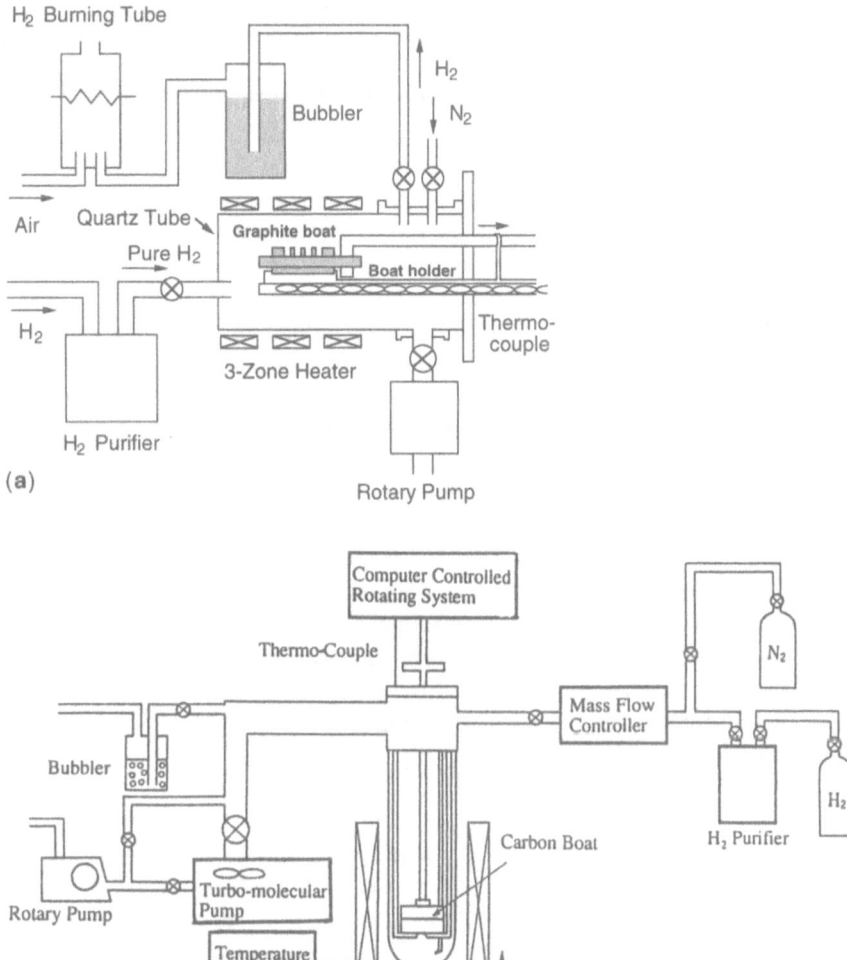

Fig.5.2. (a) Schematic representation of an LPE system with a horizontal reactor. **(b)** Schematic representation of an LPE system with a vertical reactor

heated. Metal gaskets are applied to join metals. Pure quartz is popularly employed in crystal-growth processes performed around 800°C for two reasons. First, quartz hardly discharges impurities into semiconductor materials even at high temperature, and second, the linear expansion coefficient is very small. However, it has been reported that Si resolved from the quartz mixes into semiconductor materials at a very low level.

A one-side-sealed quartz tube has become popular because it can easily introduce a movable heater system by holding only the open end of the tube. Moreover, in the case of the worst accident, the reactor exploding, the sealed end breaks first. Thus, one will be safe if one does not stand in front of or near the sealed end.

A graphite boat and a boat holder are placed inside the quartz tube, schematically shown in Fig.5.2a. The boat holder consists of a quartz saucer and a quartz bar. The thermal expansion coefficient of graphite is $\approx 3 \cdot 10^{-6}$ while that of the quartz is $\approx 5 \cdot 10^{-7}$. Thus, enough room should be kept to accommodate the size of the quartz saucer in order to avoid breaking the holder after considering the difference in the thermal expansion coefficients. For instance, a 20 cm long graphite boat at room temperature expands to 20.05 cm at 800°C while a quartz holder of the same length expands to only 20.008 cm. Figure 5.3 exhibits a bird's-eye view of the graphite boat. The graphite boat consists of an outer frame, a substrate holder, a spacer and a melt holder. The typical thickness of a substrate is 350 μm, while the typical depth of a slot on the substrate holder is precisely 400 μm, taking into account the clearance between substrate and spacer. Extremely wide clearance causes transportation of the bottom of the melt into the next, and extremely narrow clearance might deteriorate the surface morphology by contact between substrate and spacer. A fine adjustment of the clearance is made possible by examining a substrate whose thickness has been optimized by checking the surface morphology of grown wafers. The surfaces of the substrate holder and the spacer must be carefully polished to prevent multiple melts from mixing with each other, and a worn-out boat should be replaced with a new one.

Several epitaxial layers must be grown to form double-heterostructure laser diodes; for instance, a n-type cladding layer, an active layer, a p-type cladding layer and a cap layer. For this purpose, at least 4 melt boxes should be prepared in the graphite boat. Six or more melt boxes are useful to take into account a burying growth process. Each layer is successively grown by pulling the substrate holder along with the substrate from the first melt to the next. The substrate is kept at each supersaturated melt for a certain time under a fixed cooling rate, usually 0.5°C/min., according to a program, as shown in Fig.5.4.

Fig.5.3. A bird's-eye view of a graphite boat

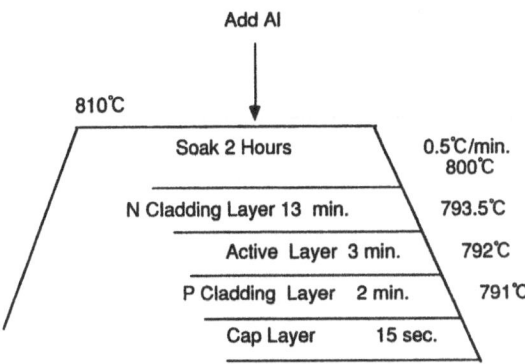

Fig.5.4. Growth program used for the growth of Surface Emitting (SE) laser wafers

In order to grow a GaAlAs/GaAs system, Al has to be separated with a Ga solution in order to prevent oxidation by the residual oxygen in the Ga melt. Other materials, especially Zn, Te and As which have a large saturated vapor pressure, can be put into Ga melts to avoid evaporation. The GaAs substrate is also covered with a GaAs piece and a graphite cover to suppress the evaporation of As until the substrate starts contacting the melts. After exposing the Ga melt to the purified H_2 for one hour at $\approx$ 800°C for out-gassing, Al is mixed with the melt by pushing a melt holder.

In order to form a uniform epitaxial layer over an entire wafer, the thickness of the melt should be uniform, too. We sometimes employ a melt-cutting mechanism, schematically shown in Fig.5.6c. A uniform thickness of the melt can be obtained by removing the top part of the melt by pushing the melt holder. By decreasing the thickness of the spacer, a very thin melt ($3 \div 5$ mm) can be obtained. This thin melt enables us to grow an ultrathin epitaxial layer such as a quantum well.

The temperature is monitored by a thermocouple such as a (Pt:87%, Rh:13%)-Pt inserted into the reactor. The (Pt:87%, Rh:13%)-Pt thermocouple is more appropriate for this kind of crystal growth than any other thermocouple because of its wide measuring range (T = −59°C to 1500°C) and its measurement accuracy. Opposite ends of the thermocouple are connected with the thermocouple extension lead wires. The voltage across the ends of the extension lead wires is measured by a voltmeter while keeping these ends at 0°C. The thermocouple should be movable and positioned along the graphite boat as in Fig.5.2. A measuring point should be located as nearly as possible to the substrate. We adjust each zone of a 3-zone heater according to the measured temperature profile.

5.2.2 Vertical Reactor

The graphite boat of a vertical reactor is illustrated in Fig.5.2b [5.1]. The vertical reactor is also explained by refering to our reactor. This system is being used for growing a GaInAsP/InP crystal. The diameter of the boat is 92 mm. The boat consists of a substrate holder and a melt holder. We do not have to prepare a melt slider for a GaInAsP/InP system. A 13×20 mm^2 slot with a depth of 400 μm is placed in a radial direction, as shown in Fig.5.2b. There are two grooves crossing one another at the bottom of the substrate holder to facilitate attachment to the boat holder while the melt holder rotates against the substrate holder. A cylindrical hole of 33 mm in diameter is made in the center of the substrate holder while a cylindrical projection of the same size exists in the melt holder. Thus, both holders are joined to one another. Seven rectangular melt boxes and a substrate preparation chamber are prepared in this particular melt holder in a revolutionary symmetry. At the beginning of the crystal growth, the slot of the substrate holder goes just under the substrate preparation chamber. The saturation pressure of phosphorus is so high that phosphorus should be prevented from escaping from the substrate. Therefore, the substrate has to be covered with an InP wafer and a graphite cover. A larger diameter graphite boat is required as the number of melt boxes increases. A rectangular hole of 10×10 mm^2 is prepared at the top center of the melt holder. A quartz bar of the same size is inserted into this hole and rotates the melt holder. It should be noted that the contact surfaces on the top of the substrate holder and on the bottom of the melt holder need to be precisely polished. The substrate makes contact from one melt to the other by rotating the melt holder. The contact time of the substrate with the melt thereby differs a little in the radial direction.

What are the necessary means to maintain strictly uniform epitaxial layers in the substrate, especially when the aim is to make very thin epitaxial layers? We can easily obtain a temperature uniformity among the melt boxes if we set the axes of a heater and a boat precisely, despite a temperature gradient occurring in a radial direction. Therefore, a simple one-zone-heater is sufficient to obtain temperature uniformity. However, a temperature gradient along the axis occurs. In other words, unlike the horizontal reactor, there is another parameter which can control the temperature gradient from the melt to the substrate by changing the temperature distribution along the axis by a vertical three-zone-heater. A thermocouple is set near to the boat, as seen in Fig.5.2b. It enables the temperature to be measured in the axial direction. The quartz tube is a one-side-sealed tube, with dimensions $\approx$70 cm long and 110 mm in diameter. Care should be taken when the boat or other things are placed inside the tube, one must avoid dropping them down to the bottom of the tube and causing it to break. The reactor is commonly designed with purified hydrogen flowing from bottom to top. In a vertical reactor, a movable heater system can easily be introduced.

Fig.5.5. Globe box for loading sources for LPE

5.3 Loading Sub-System

An outside view of a loading box for a horizontal reactor is presented in Fig. 5.5. A loading box is installed; it is filled with nitrogren gas to purge the system of air. The loading box is surrounded by a rubber buffer and placed inside a stainless steel flange in order to avoid introducing air or dust into the reactor. It is important to avoid direct inhalation of harmful gases or powders from the reactor. When we install or remove the graphite boat from the reactor, the boat on the quartz saucer is pushed in or pulled out by the quartz saucer. The quartz saucer is fixed to a truck on rails to keep the quartz bar horizontal. Thus, the length of the box is required to be at least as long as the bar. We also prepare a preloading box to prevent air from mixing into the loading box when materials are moved between the inside and the outside of the loading box. The pre-loading box can also be filled with nitrogen gas. It is convenient to build a loading platform inside the box on which the graphite boat is placed. Working space inside the box for installing or removing materials from the boat should be taken into account. Rubber gloves should easily be accessible in the lab for use in loading samples. Other factors must be taken into account when designing the size of the loading box, such as ease of material handling.

5.4 Pump and Exhaust Sub-System

In a pump and exhaust system, a rotary pump and a diffusion pump, or a turbo-molecular pump is used to pump the air out before raising the temperature. Some other oil-free pump and exhaust sub-systems, composed of either a Ti sublimation pump, a sorption pump, an ion pump and so on, have also become popular. Several kinds of pumps must sequentially be used to pump from atmospheric pressure to high vacuum. The reactor is pumped down before heating until it has reached a vacuum level of $\approx 10^{-6}$ Torr or less. The vacuum level is monitored by a Geissler tube when it is above 10^{-2} Torr and by an ionization gauge when it is below that level. (Less than 10^{-3} Torr is preferable to ensure the gauge's long lifetime). The gauge should be heated at around 10^{-4} Torr in order to degas it and to precisely measure the vacuum level. Either at high vacuum or at almost atmospheric pressure, an electric discharge does not occur in a Geissler tube. One should be aware of this in order to avoid making a mistake in determining the vacuum level. As the vacuum level increases, the color of the discharge changes from violet to transparent. At this point, the electric discharge in the Geissler tube finally stops.

The following sections explain the structure of typical vacuum pumps, such as a rotary pump, a diffusion pump and a turbo-molecular pump, as well as how to use them.

A rotary pump is one of the most fundamental pump machines, operating from atmospheric pressure to $\approx 10^{-3}$ Torr. The vapor pressure of a rotary pump is high, thus the reactor should not be pumped for a long time at high vacuum so as not to soil it with oil vapor. The rotary pump is utilized as a first-stage or subsidiary pump. When the rotary pump is turned off, do not forget to open a leak valve, otherwise oil will enter the chamber and cause terrible oil contamination.

The working range of a diffusion pump spans 10^{-2} Torr to less than 10^{-6} Torr. A diffusion pump utilizes a jet flow of oil vapor in order to pump gases from a chamber. The heater at the bottom of an oil pool vaporises the oil and causes the oil jet to flow through a small nozzle and pumps the gases out. An oil trap liquifies the useless oil vapor.

The pump and exhaust system which consists of a turbo-molecular pump and a rotary pump is a so-called *oil-free system*. The fans of the turbo-molecular pump rotate at high speed to pump the chamber, so that the rotary-pump oil hardly evacuates into the chamber nor soils it. This system requires neither a water jacket nor an oil trap.

5.5 Gas-Flow Sub-System

A schematic diagram of a gas-flow line is represented by Fig.5.13. A stainless steel pipe of 8 mm in diameter is usually utilized for this sort of system. The system should be designed to potentially reach vacuum levels of 10^{-6} Torr. Extremely purified hydrogen gas obtained through a heated Pt-Pd film should always be flowing through the reactor at high temperature ($\approx 800\,°C$) during the crystal growth in order to quickly trap any oxygen gas generated in the reactor. When we make a GaAlAs/GaAs semiconductor laser, we must especially take care of the oxidization of aluminum (Al). The Al tends to be oxidized and its oxidized layer obstructs the crystal growth,. A flow meter is useful in a LPE system to control the flow rate of purified hydrogen.

For safe exhaustion from the reactor, an anti-backflow valve and a bubbler are introduced to prevent air from re-entering the reactor. The anti-backflow valve is a mechanically directed one-way valve to the outside of the reactor; however, this valve does not work well for a quick response. Therefore, the bubbler is employed for perfect isolation and quick response. One of the bubbler bottles close to the air is filled with diffusion-pump oil. The exhausted hydrogen is released through the oil to the outside. The air mixing in the reactor is thereby perfectly cut off while the flow is kept to the outside. The bubble can also instantaneously work against a rapid inhalation into the reactor until the oil is almost inhaled into the other bottle. Meanwhile, the anti-backflow valve can assist in preventing the backflow after a little delay. An exhausting valve is inserted between the reactor and the anti-backflow valve. The exhausting valve works to keep the reactor at high vacuum while pumping the reactor, compensating for the leaking of the anti-backflow valve at high vacuum.

Other gas lines that need to be prepared include the following: a nitrogen gas line is necessary to fill the reactor and to purge air out from the loading box. When we want to do a mass-transport process for InP, an extra PH_3 gas line must be prepared, with a mass-flow controller and decomposing equipment for exhausted PH_3.

There are two methods of starting and stopping a hydrogen purifier. As mentioned before, one which we employ is to pump both an inflow and the purified sides of a Pt-Pd film until the film becomes hot or cool. The other is to flow inert gas in the inflow side in order to pump the purified side during heating or cooling. Because a Pt-Pd film is a semipermeable membrane through which only hydrogen can move, such that pure hydrogen in the purified side can diffuse only to the inflow side, introducing a vacuum in the purified side. This pumping method is economical in saving nitrogen gas. However, when a sudden accidental shutting down of the electric current is considered, an inert gas line must be installed.

5.6 Heating Sub-System

A 3-zone heater is employed to maintain an even temperature distribution along the graphite boat. Thermal uniformity and its reproducibility are the most important factors necessary to control the precise thickness of an epitaxial layer. An even temperature distribution within $\pm 0.1\,^\circ\text{C}$ is normally maintained along 20 cm. We can examine the thermal uniformity by moving a thermocouple along the graphite boat. A 3-zone heater is usually controlled by a master-slave method via changing the turn-on period of the thyristors of the power supply.

A movable heater sub-system is convenient to cool down a reactor as rapidly as possible after crystal growth. Another way to rapidly cool down the boat while the heater is fixed is to use a gate valve. By pulling the boat still in the quartz tube out of the heater through the gate valve, we cool down the boat with an electric fan. The gate valve is then closed while hydrogen is flowing in the space behind the gate valve. Next, the space in front of the gate valve is exhausted and filled with an inert gas. The remaining procedures are as previously mentioned.

5.7 Maintenance

In order to have a high-performance LPE system, one must pay special attention to maintenance. The following subsections discuss how to obtain and maintain good performance.

5.7.1 Maintenance of a Graphite Boat

A graphite boat has a certain lifespan. An old boat causes degradation of quality such as the surface morphology or the nominal threshold current density of an epitaxial wafer. If the quality of the wafer degrades, we should change to a new boat (typically after $150 \div 200$ runs). A new boat is usually washed in pure water with an ultrasonic cleaner for half a day or a day at the beginning. After drying the washed boat we set it in the reactor.

5.7.2 Baking of the Reactor

The inside of a reactor is contaminated with dopants after several crystal-growth sessions. We need to bake the reactor every 2 or 3 times a crystal is grown, at least. For example, our lab bakes the reactor at 900°C for 8 hours in under vacuum condition after two crystal have been grown.

5.8 Liquid-Phase Epitaxy

There are roughly four techniques for growing epitaxial layers by the LPE method, each of which is different thermodynamically. The following techniques will be discussed in detail: step-cooling, equilibrium-cooling, supercooling, and two-phase solution.

● In the step-cooling technique, the substrate comes into contact with the melt just after it is cooled by a ΔT lower than the saturation temperature. The substrate remains fixed at this temperature for a certain period of time.
● The equilibrium-cooling technique is used to grow material from a saturated melt by decreasing the temperature at a fixed cooling rate. The substrate and melt are kept in equilibrium.
● The supercooling technique is a combination of the step-cooling and equilibrium-cooling techniques. The substrate is drawn under a solution, and when the temperature of the melt is lowered by ΔT from the saturation temperature, it is cooled down at a fixed cooling rate. The supercooling technique is similar to step cooling before the melt reaches saturation, and then it approached the equilibrium-cooling technique.
● Finally, in the two-phase solution technique, the melt is sufficiently supersaturated. Spontaneous nucleations occur in the melt. The supersaturation is then maintained during the crystal growth. Since crystal grows not only on the substrate but also on the seed, the crystal growth on the substrate is rather slow. This technique is suitable for the epitaxial growth of thin layers, $0.05 \div 0.1$ μm in thickness. Either the step-cooling or supercooling technique is employed for the GaAlAs/GaAs system.

The two-phase solution technique is very popular, especially for the GaInAsP/InP system. The phosphorus in the melt, which is likely to have been vaporized, can be saturated. The thickness dependence of each technique on the growth time has been thoroughly studied for the GaAlAs/GaAs system and it is in good agreement with the experimental results [5.2]. The growth thickness is expressed for each case as

Step-cooling technique

$$d = 2\Delta T \frac{\sqrt{(D/\pi)t}}{C_s m} . \qquad (5.1)$$

Equilibrium-cooling technique

$$d = \frac{4}{3}R\,\frac{\sqrt{(D/\pi)t^3}}{C_s m}\,.$$ (5.2)

Supercooling technique

$$d = \sqrt{\frac{D}{\pi}}\,\frac{2\Delta T t^{1/2} + (4/3)R t^{3/2}}{C_s m}\,.$$ (5.3)

Here $m = \dfrac{\Delta T}{C_0^l - C_s^l}$ (5.4)

and the symbols used in (5.1-4) are as summarized:

d   Thickness of the epitaxial layer
t   Growth time
T_1 Saturation temperature
ΔT Supersaturation degree which is the temperature difference from the saturation temperature
R   Cooling rate
C_2 Concentration of As in a GaAs solid
D   Diffusion constant of As in the solution
C_0^l Concentration of As at T_1 for $t = 0$
C_s^l Concentration of As on the liquidus curve at $T_1 - \Delta T$.

Figure 5.6 presents the calculated result of each epitaxial growth technique for the following parameters [5.2]:

$T_1 = 800°C$, $m = 4.0 \cdot 10^3$ K(atom fraction)$^{-1}$,
$D = 4 \cdot 10^{-5}$ cm$^2 \cdot$ s^{-1} ,
$\Delta = 5°C$, $R = 0.6°C/min$,
$C_s = 1/2$ (atom fraction) . (5.5)

The step-cooling technique is superior in terms of its thickness control, because the growth rate is low, being proportional to $t^{1/2}$. At the beginning of crystal growth, the growth speed is proportional to $t^{1/2}$ in the supercooling technique. As the growth time increases, the time dependency of the growth speed changes from $t^{1/2}$ to $t^{3/2}$. This means that the supercooling technique is excellent for a wide range of thicknesses from submicrometer to several tens of micrometers. Equation (5.3) becomes valid also for the GaInAsP/InP system by multiplying the right side of the equation by 1.94, a

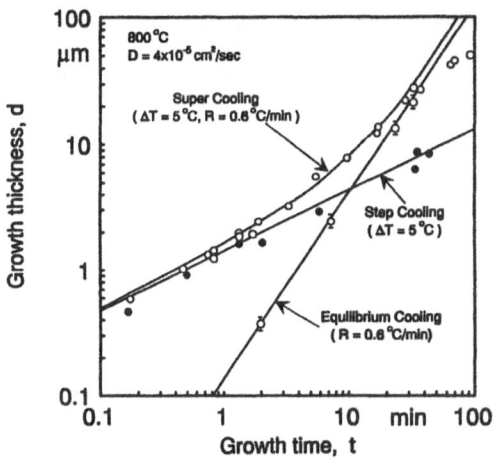

Fig.5.6. Growth thickness for several cooling techniques [5.2]

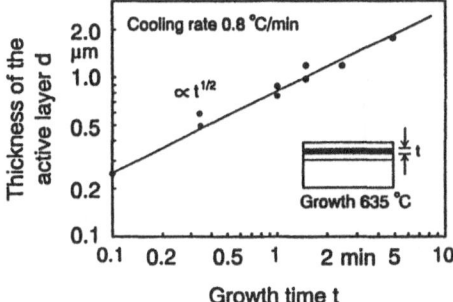

Fig.5.7. Relationship between growth time and thickness of 1.3 μm GaInAsP layer [5.4]

correction coefficient [5.5]. As shown in Fig.5.7, the active layer thickness is almost proportional to the square root of the growth time ($t^{1/2}$) within 5 min for GaInAsP/InP growth.

5.9 LPE Process

5.9.1 GaAlAs/GaAs System

a) Determination of the Source-Material Quantity

We can theoretically estimate the equilibrium weights x of GaAs and y of Al per 1 gram of Ga melt from the following equations.

$$\frac{x}{M_{Al}} = X_{Al}^{\ell} \left[\frac{2y}{M_{GaAs}} + \frac{x}{M_{Al}} + \frac{1}{M_{Ga}} \right],$$ (5.6)

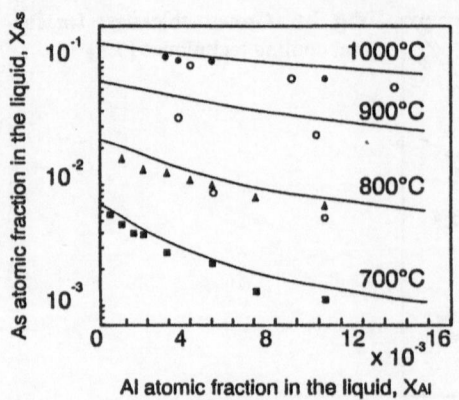

Fig.5.8. Liquids isotherms in the AlGaAs system [5.5, 6]

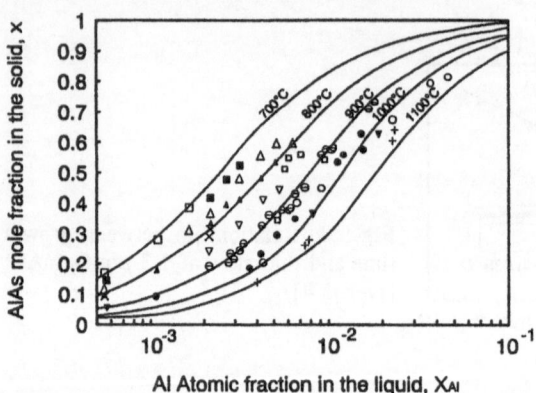

Fig.5.9. Solidus compositions in $Al_xGa_{1-x}As$ as function of liquid's composition [5.5, 6]

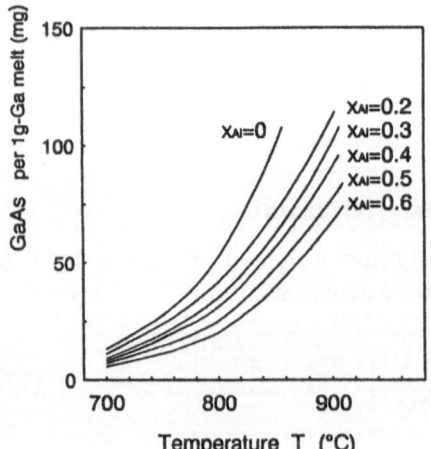

Fig.5.10. Saturated weight of GaAs per 1 g Ga melt

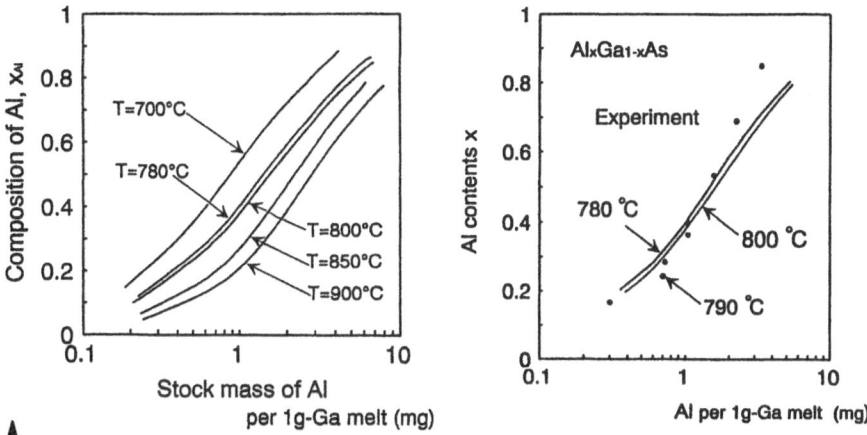

Fig.5.11. Al content X_{Al} of $Ga_{1-x}Al_xAs$ versus Al weight per (1–g) Ga melt

Fig.5.12. Measured results of the Al content x versus Al weight per (1–g) Ga melt

$$\frac{y}{M_{GaAs}} = X_{As}^{\ell}\left[\frac{2y}{M_{GaAs}} + \frac{x}{M_{Al}} + \frac{1}{M_{Ga}}\right], \tag{5.7}$$

here X_{Al}^{ℓ} and X_{As}^{ℓ} are the atomic fractions of Al and As in the liquid, and M denotes the molecular weight. First X_{Al}^{ℓ} is determined at the desired Al content X_{Al} in the solid from Fig.5.8. Next, X_{As}^{ℓ} is extracted from Fig.5.9. We arrange the results to the diagrams shown in Figs.5.10,11.

Since the monitoring point for the temperature is different among apparatuses, the saturation weight of GaAs per 1 gram of Ga melt at a certain temperature should be measured as follows. A GaAs substrate whose weight has been measured already, is melted back into the pure Ga melt until equilibrium occurs at a certain temperature. The difference of the GaAs substrate between before and after being melted back can tell us the saturation weight of GaAs per 1 gram of Ga melt at a particular temperature. In the same manner, the saturation weight of GaAs in Ga-rich plus Al melt for a $Ga_{1-x}Al_xAs$ layer can be measured. Based on these results, the supersaturation degree ΔT, the key factor of the supercooling technique, can be decided. As a concrete example, ΔT is +4°C for GaAs or +7°C for GaAlAs. It is desirable to apply enough ΔT to the epitaxial growth of GaAlAs, because ΔT is very sensitive to the variation of the Al weight in a Ga-rich melt. Thus, the supersaturation degree +7°C might be preferable to +4°C for GaAlAs.

The Al weight per 1g-Ga for $Ga_{1-x}Al_xAs$ has been confirmed by measuring the lattice constant of the $Ga_{1-x}Al_xAs$ layer by the X-ray diffraction method. The details are described in Sect.7.4. The results measured in our experiments and those of *Panish* and *Hayashi* [5.5] are detailed in Fig.5.12.

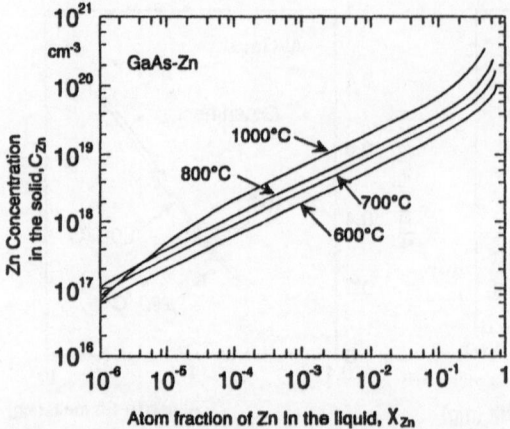

Fig.5.13. The Zn concentration in GaAs versus the atom fraction of Zn in the liquid along the 600°, 700°, 800°, and 1000°C Ga-rich liquid isotherms in the Ga-As-Zn system [5.7]

The doping concentration of a GaAs epitaxial layer can be adjusted by changing the dopant weight. Figures 5.13 and 14 list the atomic fraction of Zn, Ge and Sn in the liquid. With (5.6, 7), the dopant weight z per 1 gram of Ga melt can be obtained in the same way, as mentioned above, i.e.,

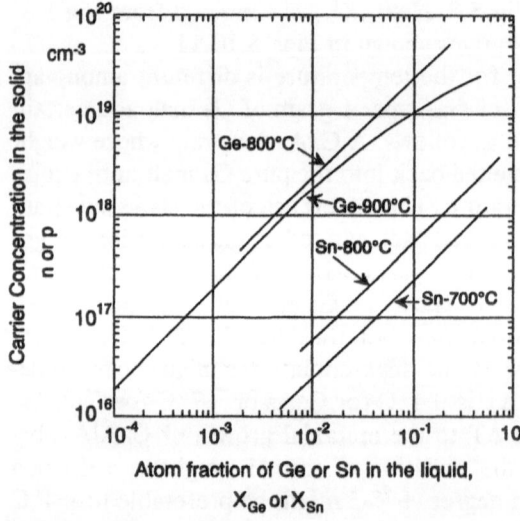

Fig.5.14. The upper curves are the room-temperature hole concentration in GaAs versus the atomic fraction of Ge in the liquid along the 800° and 900°C Ga-rich liquidus isotherms. The lower curves are the room-temperature electron concentration in GaAs versus the atomic fraction of Sn in the liquid along the 700°C and 900°C Ga-rich liquidus isotherms [5.6, 8-10]

$$\frac{z}{M_d} = X_d^\ell \left[\frac{2y}{M_{GaAs}} + \frac{z}{M_d} + \frac{1}{M_{Ga}} \right],$$ (5.8)

The Te weight per 1 gram of Ga melt at a certain doping concentration can be determined taking advantage of Fig.5.15a,b in the following manner. First, a certain distribution coefficient, being the ratio of the distribution coefficient at a growth temperature equal to that at 1000°C, is estimated from Fig.5.15a. This figure indicates that the higher doping concentration in the solid is obtained at a lower growth temperature even though the same weight of Te is in the melt. Figure 5.15b indicates the relationship between the Te concentration C_{Te} in the solid versus the atomic fraction in the liquid X_{Te}^ℓ at 1000°C. Thus, this value must be corrected by the distribution coefficient which was obtained from Fig.5.15a.

The aforementioned calculations can be understood further by considering the carrier concentration C_{Te} to be 10^{18} cm^{-3} in a GaAs epitaxial layer at the growth temperature $T_g = 800$°C. Figure 5.15a yields the distribution coefficient 6.4. Therefore, we read the value of X_{Te}^ℓ at $C_{Te} = 10^{18}/5.4 = 1.1 \cdot 10^{17}$ cm^{-3} in Fig.5.15b.

For the doping of a GaAlAs layer, we have to take into account the decrease of the doping concentration with the increase of the Al content in the epitaxial layer, depicted in Figs.5.16. The doping concentration should be verified by a Hall measurement or a capacitance-voltage (C-V) measurement, as described in Sect.7.5.

b) LPE Procedure

The LPE procedure is explained below with reference to the sequence in the flow chart illustrated in Fig.5.17.

(i) *Weighing*. An example of weighing materials for a DH laser is shown in Table 5.1. It is particularly important to measure dopants such as Al or Te metals very precisely.

(ii) *Rinse and etch of the source materials*. According to the recipe in Table 5.2, the source materials are rinsed and etched in order to clean the surface. Dopants such as Ge and Zn should be cleaned before weighing.

(iii) *Preparation*. After drying the etched materials, they are loaded into each melt box of the graphite boat. The Al metal is separated from the other materials, as mentioned in Sect.5.2.

(iv) *Crystal growth*. A typical growth program is illustrated in Fig.5.30, which shows GaAlAs/GaAs wafers for surface emitting lasers. First, Ga-rich melts, including GaAs and Ge or Te, are baked in a purified hydrogen ambiance at 810°C for an hour to drive out the oxygen gas. Next, Al solutes and Ga-rich melts are mixed at 810°C for an hour. Then, the LPE system reactor begins cooling at the rate of 0.5°C/min. According to the schedule

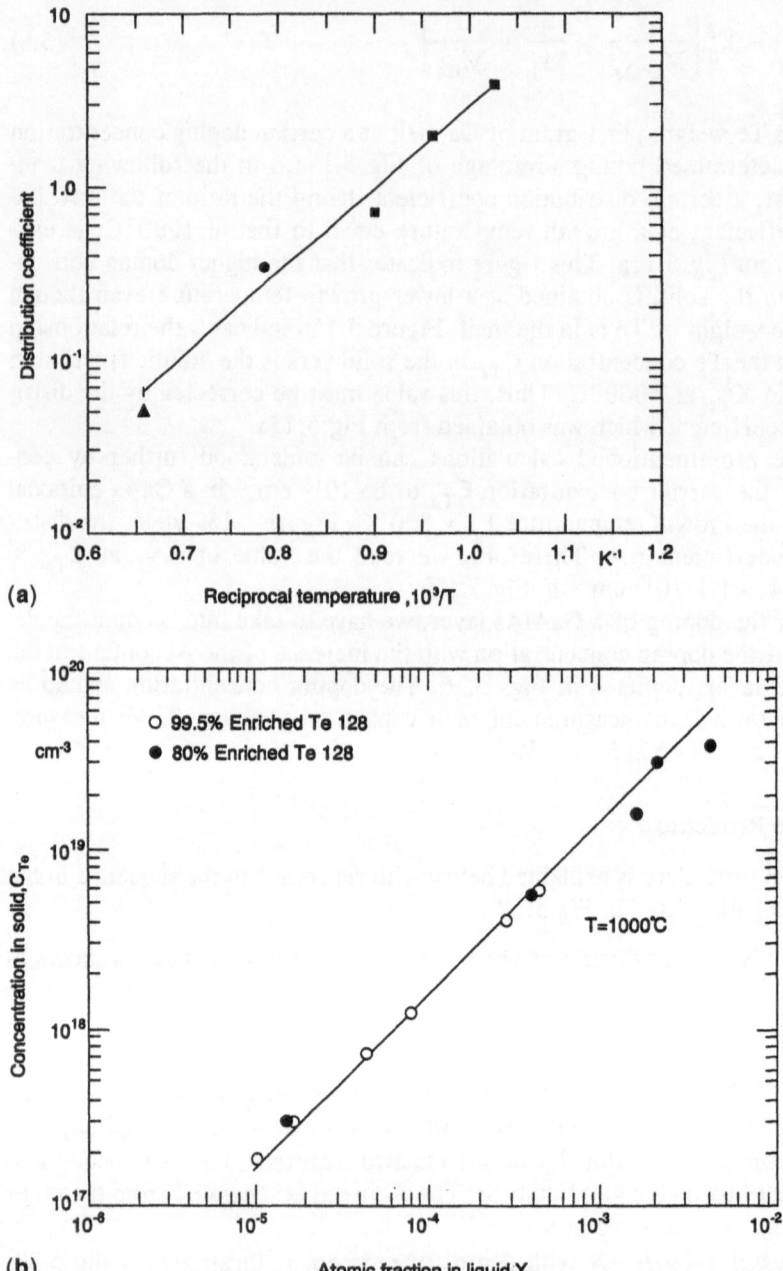

Fig.5.15. (a) The distribution coefficient of Te in GaAs as a function of temperature [5.11-13]. (b) A portion of the 100°C solid-solubility isotherm Te in GaAs [5.11, 14]

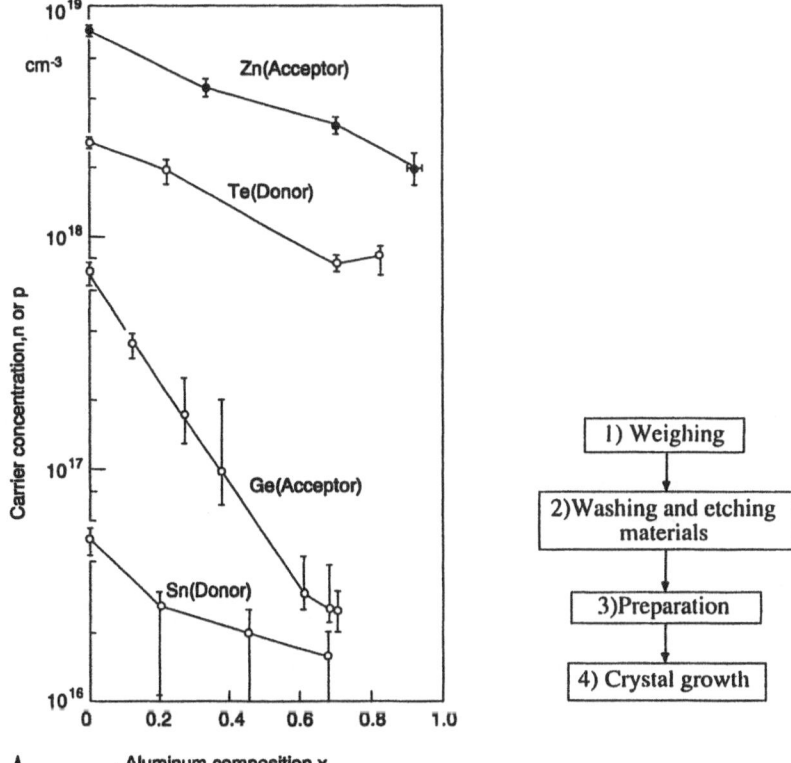

Fig.5.16. Carrier concentration at 300 K in $Al_x Ga_{1-x} As$ for four dopants. The dopant mole percent in the solution used to grow the layers by LPE was 0.5 % for each point. Layers were grown on (100) GaAs at temperatures of $840 \div 810°C$ [5.15]

Fig.5.17. Flow chart of the LPE procedure

shown in Fig.5.4, the GaAs substrate is driven to the first melt and then to the others.

5.9.2 GaInAsP/InP System

As mentioned in Chap.2, $Ga_x In_{1-x} As_y P_{1-y}$ contains two controllable parameters, enabling independent adjustment of the lattice constant and the band-gap energy. The lattice constant a(x, y) of $Ga_x In_{1-x} As_y P_{1-y}$ is given as follows [5.16, 17]:

$$a(x, y) = xy \cdot a(GaAs) + x(1-y) \cdot a(GaP) + (1-x)y \cdot a(InAs) + (1-x)(1-y) \cdot a(InP). \tag{5.9}$$

Table 5.1. Weighing example of a DH laser

	N-clad layer	Active layer	P-clad layer	Cap layer
Degree of super saturation ΔT [°C]	7	4	7	4
Growth temperature [°C]	800	793.5	792	791.7
Material weight per 1g-Ga melt				
GaAs [mg]	35	45	32.5	41.5
Al [mg]	0.73	0	0.69	0.2
X_{Al}	0.3	0	0.3	0.1
Dopants weight per 5g-Ga melt				
Te [mg]	0.15			
[cm^{-1}]	$1 \cdot 10^{18}$			
Ge [mg]		7	90	150
[cm^{-1}]		$4 \cdot 10^{17}$	$5 \cdot 10^{17}$	$3 \cdot 18^{18}$

Table 5.2. Etching procedure of each solute

Material	GaAs	Al	Ge	Te
Ultrasonic cleaning	CH_3OH 5 min $Cl\text{-}CH\text{=}CCl_2$ 10 min CH_3OH 5 min			
Etchant	$Br + CH_3OH$	$HCl:H_2O$ $= 1:1$	$HF:HNO_3$ $= 1:1$	$HCl:H_2O$ $= 1:3$
Etching time	3 min	2 min	2 min	1 min
Etch stop	CH_3OH	H_2O	H_2O	H_2O
Solution substitution	CH_3OH	H_2O $\downarrow$ CH_3OH	H_2O $\downarrow$ CH_3OH	H_2O $\downarrow$ CH_3OH

According to the results measured by *Nahory* et al. [5.19], the binary lattice constants are: a(GaAs) = 5.653 Å, a(GaP) = 5.4512 Å, a(InAs) = 6.0590 Å, and a(InP) = 5.8696 Å. Equation (5.10) is obtained by inserting these data into (5.9)

$$a(x,y) = 0.1894y - 0.4184x + 0.013xy + 5.8696 \text{ [Å]} . \tag{5.10}$$

The relation between x and y, therefore, is given by the following equation, when the a(x, y) coincides with the lattice constant of InP:

$$0.1894y - 0.4184x + 0.0130xy = 0 . \tag{5.11}$$

Usually, (5.11) is approximated as

$$x = 0.467y . \tag{5.12}$$

Next, the relation between the band-gap energy and compositions of x and y based on the theory of *Moon* et al. [5.16] as well as newly obtained data [5.18] is given by (5.13):

$$\begin{aligned} E_g(x,y) = {} & 1.35 + 0.672x - 1.091y + 0.758x^2 + 0.101y^2 \\ & - 0.157xy - 0.312x^2y + 0.109xy^2 . \end{aligned} \tag{5.13}$$

The band-gap energy calculated in terms of x and y which satisfies (5.11) has been confirmed to almost coincide with the phenomenological relation of *Nahory* et al. It is expressed by [5.17]

$$E_g(y) = 1.35 - 0.72 + 0.12y^2 \tag{5.14}$$

and is depicted in Fig.5.18.

A three-dimensionally displayed band-gap energy versus compositions of x and y is illustrated in Fig.5.19. With the aid of this figure, from a contour line of the band-gap energy versus the compositions of x and y, the band structure of GaInAsP lattice-matched to InP is obtained for the entire allowed composition of y. In contrast, the band gap of GaInAsP in the vicinity of GaP is indirect.

Next, let us discuss how to determine the composition of an In-rich melt which exists in equilibrium with the solid phase of $Ga_x In_{1-x} As_y P_{1-y}$ at the desired compositions of x and y. Calculation of its phase equilibrium was described by *Yamamoto* [5.18]. A more complicated derivation is necessary, but only the main results are summarized here due to space limitations. Figure 5.20 illustrates the relation between the atomic fraction of Ga in the liquid X_{Ga}^ℓ and the Ga composition x in the solid at equilibrium. Figure 5.21 delineates the same relation between the atomic fraction As in

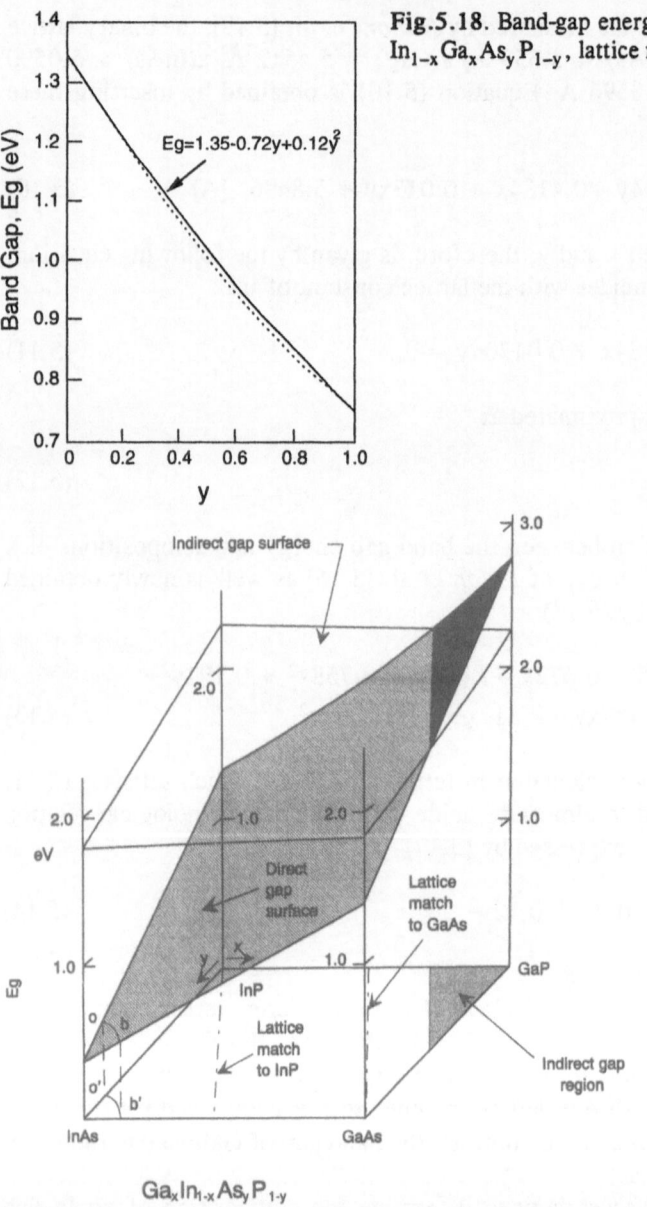

Fig.5.18. Band-gap energy versus content y of $In_{1-x}Ga_xAs_yP_{1-y}$, lattice matched to InP

$Eg=1.35-0.72y+0.12y^2$

$Ga_xIn_{1-x}As_yP_{1-y}$

Fig.5.19. Band-gap energy diagram of $In_{1-x}Ga_xAs_yP_{1-y}$ at 300 K [5.19]

the liquid X_{As}^ℓ and the As composition y. Figure 5.21 shows the X_P^ℓ dependence on X_{Ga}^ℓ when the In-rich solution is in equilibrium with the $Ga_xIn_{1-x}As_yP_{1-y}$ solid.

The As composition y in the $Ga_xIn_{1-x}As_yP_{1-y}$ solid of the desired band-gap energy is given by (5.14) while keeping the lattice constant equal

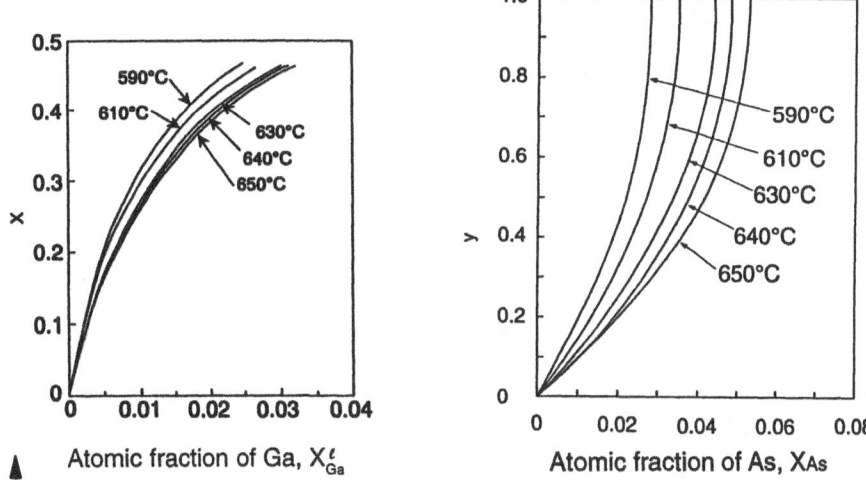

Fig.5.20. Composition x of $In_{1-x}Ga_xAs_yP_{1-y}$ versus atomic fraction of Ga at equilibrium [5.18]

Fig.5.21. Composition y of $In_{1-x}Ga_xAs_yP_{1-y}$ versus atomic fraction of As at equilibrium [5.18]

to that of InP. The Ga composition x is obtained by (5.11 or 12). Now, the atomic fractions X_{Ga}^ℓ, X_{As}^ℓ, and X_P^ℓ in the In-rich melt which exists in equilibrium with the desired $Ga_xIn_{1-x}As_yP_{1-y}$ solid can be obtained from Figs. 5.21,22. The actual weights x, y and z of InP, InAs, and GaAs per 1 gram of In can be estimated by inserting the obtained atomic fractions into (5.15-17):

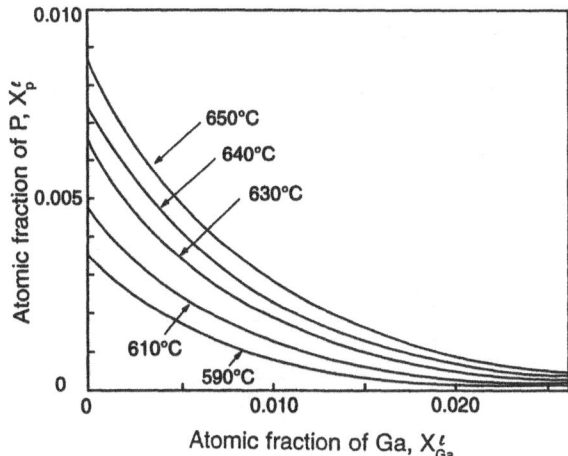

Fig.5.22. Atomic fraction of P versus atomic fraction of Ga in $In_{1-x}Ga_xAs_yP_{1-y}$ at equilibrium [5.18]

$$\frac{x}{M_{InP}} = X_P^\ell \left[\frac{2x}{M_{InP}} + \frac{2y}{M_{InAs}} + \frac{2z}{M_{GaAs}} + \frac{1}{M_{In}} \right], \tag{5.15}$$

$$\frac{y}{M_{InAs}} + \frac{z}{M_{GaAs}} = X_{As}^\ell \left[\frac{2x}{M_{InP}} + \frac{2y}{M_{InAs}} + \frac{2z}{M_{GaAs}} + \frac{1}{M_{In}} \right], \tag{5.16}$$

$$\frac{z}{M_{GaAs}} = X_{Ga}^\ell \left[\frac{2x}{M_{InP}} + \frac{2y}{M_{InAs}} + \frac{2z}{M_{GaAs}} + \frac{1}{M_{In}} \right], \tag{5.17}$$

where M_{AB} represents the molecular weight of a binary compound. The weight of each material per one gram of In is thereby determined for the desired $Ga_x In_{1-x} As_y P_{1-y}$.

As mentioned in Sect.5.9.1, the monitoring point of the temperature differs between LPE apparatuses. This necessitates scanning the actual material weight per one gram of In by trial and error with respect to the theoretical value. The lattice matching $|\Delta a/a|$ is examined by X-ray diffraction and it should be less than 0.05%. The details are described in Sect.7.4.

When we employ the two-phase-solution technique in order to always keep phosphorus in a saturation condition, the InP material should be prepared so as to be saturated. The Ga weight is adjusted to obtain the desired band-gap energy; and the As weight is made to coincide with the lattice constant of the InP substrate. Figures 5.23-26 exhibit the experimental data of material weights for the two-phase-solution technique [5.20, 22].

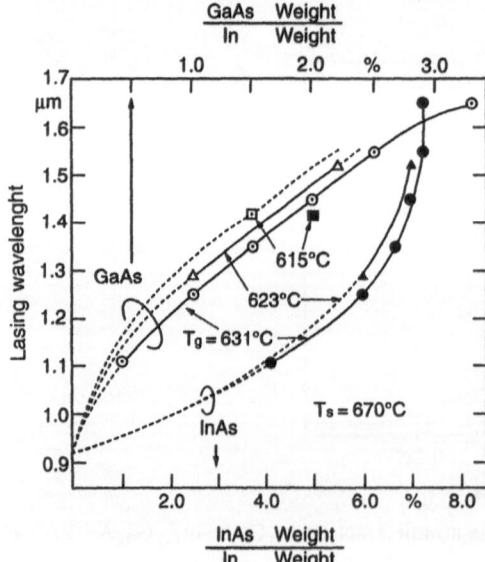

Fig.5.23. GaAs and InAs weights to In to keep the lattice match for the desired wavelength. (the weight of InP is locked to 10 mg/[(1−g) In] for the two-phase-solution technique) [5.20]

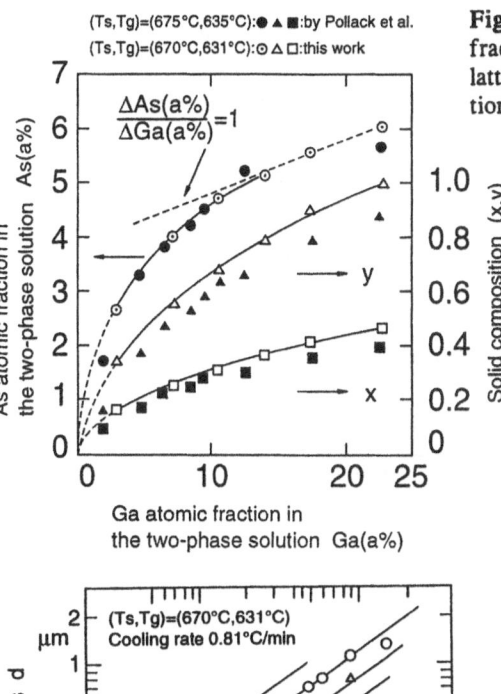

Fig.5.24. Relationship between atom fractions of As and Ga to keep the lattice match to InP at two-phase solution technique [5.20]

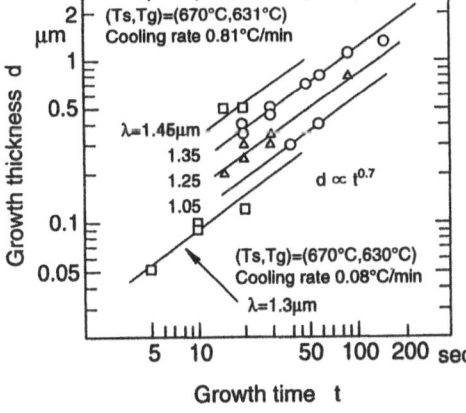

Fig.5.25. Grown thickness versus growth time in two-phase solution technique [5.20]

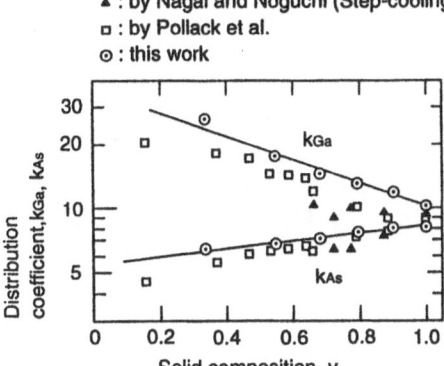

Fig.5.26. Distribution coefficient vs. composition y of $In_{1-x}Ga_xAs_yP_{1-y}$ in the two-phase-solution technique [5.20-22]

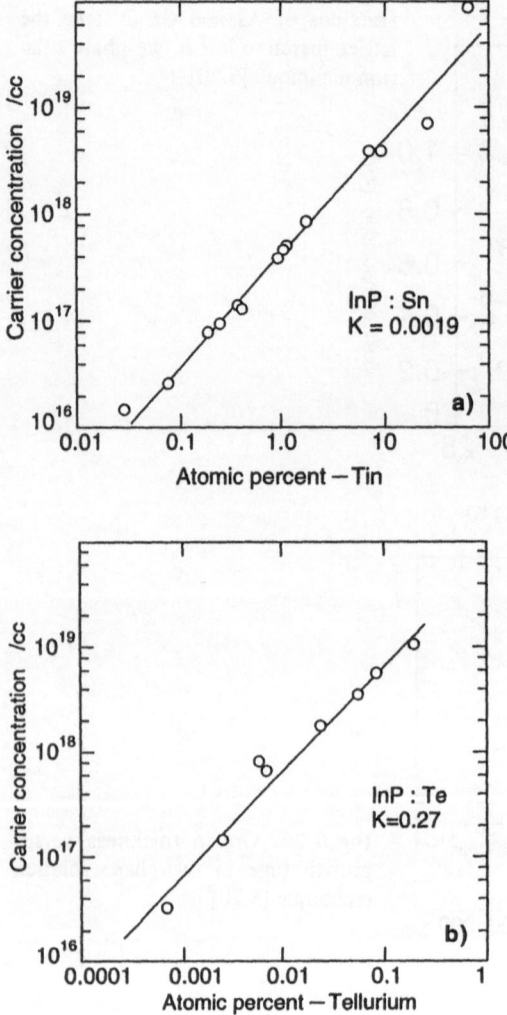

Fig.5.27. (a) Carrier concentration in an InP solid versus atomic percent of Sn in InP liquid. (b) Carrier concentration in an InP solid versus atomic percent of Te in InP liquid.

The relationship between carrier concentration and dopant weight has not been made clear. Figures 5.27a-d show the experimental relationship between the carrier concentration in the solid versus the dopant weight in atomic percent in the liquid. Figure 5.28 and Tables 5.3,4 illustrate the LPE growth process.

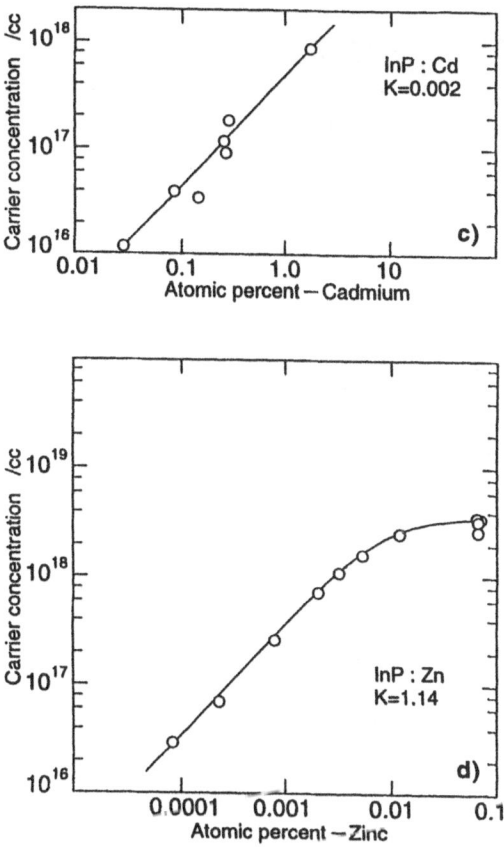

Fig.5.27. (c) Carrier concentration in an InP solid versus atomic percent of Cd in InP liquid. **(d)** Carrier concentration in an InP solid versus atom percent of Zn in an InP liquid. [5.23]

5.9.3 Other Materials

The GaAlAs/GaAs and GaInAsP/InP systems mentioned above are the most popular and valuable materials for opto-electronics due to their capability to be used as materials for both optical and electric devices. However, these two materials cannot meet the demand for light-emitting sources of the various wavelengths required, hence many other materials are being studied, and are depicted in Fig.2.3. Some materials have been investigated as possible visible-light sources and others may be useful for light sources for fluoride-glass-fiber communication systems [5.24-26] in the range of $2 \div 4$ μm or for longer wavelengths ($\lambda > 2\mu$m) light sources. This section examines several materials which are intensely studied, with the exception of GaAlAs/GaAs and GaInAsP/InP systems.

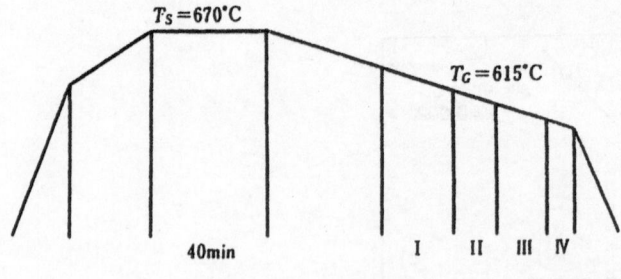

Fig.5.28. Temperature program for DH wafer ($\lambda = 1.3\,\mu$m) in two-phase-solution technique

Table 5.3. Quantities of sources for epitaxy

No	Laser	Content [mg]				
		InP	GaAs	InAs	Te/In	Zn/In
I	InP(buffer)	50			80	
II	GaInAsP(active)	50	65	225		
III	p-InP(cladding)	50				5
IV	p$^+$-GaInAsP(cap) for 5g of In solvent	50	30	155		30

Te/In: 2.5% Te in In
Zn/In: 1% Zn in In

Table 5.4 Etching procedure of each solute

Part a

Material	GaAs, InAs, InP	In	InZn, InTe
Ultrasonic cleaning	CH_3OH $Cl\text{-}CH=CCl_2$ CH_3OH	5 min 10 min 5 min	
Etchant	0.3%Br-CH_3OH	HNO_3 for 2 min H_2O:HNO_3 = 1:1 for 1 min	H_2O:HNO_3 = 1:1
Etching time	2 min		0.5 min
Etch stop solution	CH_3OH	H_2O	H_2O
Substitution	CH_3OH	H_2O ↓ CH_3OH	H_2O ↓ CH_3OH

Part b

Material	InP substrate	
Altrosonic cleaning	CH_3OH $Cl\text{-}CH=CCl_2$ CH_3OH	5 min 10 m 5 min
Etchant	H_2O:H_2SO_4:H_2O_2 = 1:3:1	
Etching time	1 min at 70°C	
Etch stop solution	H_2O	
Substitution	CH_3OH	

[a] The surface of the substrate should be mirrorlike after etching

a) Visible-Light Semiconductor Lasers

The attractive features of visible-light semiconductor lasers were already stated in Sect.2.2. This section explains the systems of InGaP/ InGaAsP and GaAlAs/InGaAsP on GaAs substrates which have been studied with the LPE system.

First, let us review the relationship between the band-gap energy and the lattice constant. Figure 5.24 shows the band-gap energy of GaInAsP whose lattice matches that of the GaAs substrate by mixing GaAs and $Ga_{0.51}In_{.49}P$ at the rate x, namely $Ga_{1-x}(Ga_{0.51}In_{0.49})_xAs_{1-x}P_x$. The other choice is the $(Al_xGa_{1-x})_{0.51}In_{0.49}P$ system with which the shortest wavelength of III-V compound semiconductor lasers can be achieved. DH structures that are lattice matched to GaAs are summarized in Table 5.5. The upper three systems of the four listed in Table 5.5 were grown with LPE, while the last one was grown with a MOCVD system.

A $Ga_{1-x}Al_xAs/Ga_{1-y}Al_yAs$ (y >x) DH structure on a GaAs substrate has a larger compressive stress at a higher Al composition due to the difference of the thermal expansion coefficient between GaAs substrate and the $Ga_{1-x}Al_xAs$ active layer. *Yamamoto* et al. [5.27] was successful with a room-temperature CW operation at 683 nm after growing a very thick $Ga_{0.8}Al_{0.2}As$ buffer layer (100 μm) and then polishing the GaAs substrate to a 4 μm thickness. The Al compositions of the DH layers are: cladding layers (y= 0.75) and the active layer (x = 0.28 ÷ 0.31). A problem with this system is that the threshold current abruptly increases being proportional to the Al content due to the decrease of the energy difference between the direct and indirect conduction minima in the active layer [5.27].

$Ga_{1-0.49x}In_{0.49x}As_{1-x}P_x/Ga_{1-y}Al_yAs$ and $Ga_{1-0.49x}In_{0.49x}As_{1-x}P_x/Ga_{0.51}In_{0.49}P$ systems on GaAs can also be grown using LPE [5.30,31]. The shortest wavelength attainable with a $Ga_{0.51}In_{0.49}P$ active layer is 660 nm. The hetero-barrier height of the $Ga_{1-0.49x}In_{0.49x}As_{1-x}P_x/Ga_{1-y}Al_yAs$ system can be assumed more than 0.2 eV at 600 nm, although that of the $Ga_{1-0.49x}In_{0.49x}As_{1-x}P_x/Ga_{0.51}In_{0.49}P$ system is less than 0.2 eV. *Kishino* et

Table 5.5. Various systems of visible light semiconductor lasers

Active	Cladding	Wavelength [μm]	
$Ga_{1-x}Al_xAs$	$Ga_{1-y}Al_yAs$ (y >x)	0.67 ÷ 0.87	[5.24]
$Ga_{1-0.49x}In_{0.49x}As_{1-x}P_x$	$Ga_{1-y}Al_yAs$	0.66 ÷ 0.8	[5.37-39]
$Ga_{1-0.49x}In_{0.49x}As_{1-x}P_x$	$Ga_{0.51}In_{0.49}P$	0.66 < ≈0.8	[5.40-43]
$(Al_xGa_{1-x})_{0.51}In_{0.49}P$	$(Al_yG_{1-y})_{0.51}In_{0.49}P$ (y >x)	0.58 ÷ 0.66	[5.44-47]

al. [5.29] reported laser oscillation in this system at 670 nm. The same fabrication procedure used for GaAlAs/GaAs and GaInAsP/InP systems could be used. An important point to note is the mixture of phosphorus into the melt aside from the one for the active layer. *Kishino* et al. [5.31] covered the melt of the active layer with a graphite cap so as not to mix the evaporated phosphorus into neighboring melts. His group obtained a threshold current density of 5.6 kA/cm^2 at room-temperature, pulsed conditions. Several groups have demonstrated laser oscillation at 705 to 810 nm [5.32-35]. The two-phase-solution technique might be useful in this system as it is in the GaInAsP/InP system in order to maintain the concentration of phosphorus in the melt. *Kawanishi* et al. [5.36] reported a threshold current density of 8 kA/cm^2 at 720 nm.

The shortest wavelength around 580 nm has been achieved by the $(Al_xGa_{1-x})_{0.51}In_{0.49}P$ system. *Kawamura* et al. [5.37] realized room-temperature operation at ≈ 660 nm. Lasing operation at 626 nm was reported by *Kobayashi* et al. [5.39, 40] using MOCVD. Other groups [5.39, 40] also realized laser operation with the same system. However, it is difficult to fabricate $(Al_xGa_{1-x})_{0.51}In_{0.49}P$ lasers using LPE.

b) Longer-Wavelength ($\lambda > 2\,\mu m$) Semiconductor Lasers

Fluoride-glass fibers are promising for long-distance optical-fiber communication due to their very low loss, one or two orders of magnitude lower than those of silica fibers at 1.6 μm. Semiconductor lasers operating at wavelengths longer than 1.6 μm are urgently needed not only for optical-fiber communication, but also for applications such as molecular spectroscopy, utilizing the wide range of lasing wavelengths available from semiconductor lasers. The $Ga_{1-x}In_xAs_ySb_{1-y}/Al_{1-x}Ga_xAs_ySb_{1-y}$ system for 2 to 4 μm [5.41- 43], the ternary alloys $Pb_{1-x}Sn_xTe$, $Pb_{1-x}Se_x$, $Pb_{1-x}Sn_xSe$, and the quaternary material system of $Pb_{1-x}Sn_xSe_yTe_{1-y}$ for $3 \div 34$ μm [5.44] have been studied in order to obtain longer-wavelength semiconductor lasers. The $Pb_{1-x}Sn_xTe$ system fabricated by LPE on PbTe substrates are especially being studied [5.45-48]. A threshold current density as low as 1.7 kA/cm^2 at room temperature was obtained in the system of $Ga_{1-x}In_xAs_ySb_{1-y}/Al_{1-x}Ga_xAs_ySb_{1-y}$ [5.49]. *Tomasetta* and *Fonstad* [5.47] reported that a threshold current of 1.2 kA/cm^2 at 77 K had been obtained with double-heterostructure lasers consisting of a $Pn_{0.82}Sn_{0.18}Te/PbTe$ system grown by means of LPE p-type PbTe substrates.

6. Vapor Phase and Beam Epitaxies

The Liquid-Phase Epitaxy (LPE) growth technique is based on simple principles and resrs upon simple apparatuses. It is described in Chap. 5. Recently, superlattice devices composed of finer heterostructures have attracted the attention of device researchers because the quantum effects can drastically improve the device characteristics. Fine-structured epitaxy for producing superlattice devices requires thickness controllability of less than several hundred Å, uniformity and heterointerface abruptness. Vapor phase and beam epitaxial growth techniques for fine-structured devices are being intensely studied. This chapter introduces and explains Metal-Organic Chemical Vapor Deposition (MOCVD) and Chemical-Beam Epitaxy (CBE).

6.1 Metal-Organic Chemical Vapor Deposition (MOCVD)

6.1.1 MOCVD System

This section describes the details of a MOCVD system in reference to the system used in our laboratory, which is designed for GaAlAs/GaAs electro-optic devices. MOCVD [6.1-3] utilizes a metal-organic gas as a material gas. Elements that are required for crystals are provided through the resolution of the metal-organic gas.

When growing electro-optic devices such as surface-emitting lasers described in Chap. 10, it is important to achieve a smooth surface on the grown layers, and to grow GaAlAs with a high aluminum content for multilayer reflectors.

Oxygen traces in the reactor chamber degrade the surface morphology and the quality of a GaAlAs crystal. Additionally, the gas-flow conditions influence the surface morphology. Therefore, a MOCVD system must be designed to eliminate system leaks and to obtain a smooth gas flow in the reactor chamber.

Figures 6.1 and 2 present a schematic diagram and a photograph of the growth apparatus, respectively. The reactor gas manifold and the mixing system are constructed using only stainless steel. The reactor consists of a single vertical quartz tube (inner diameter: $4 \div 5$ cm in the substrate area) with a watercooling jacket and a SiC-coated graphite susceptor heated in-

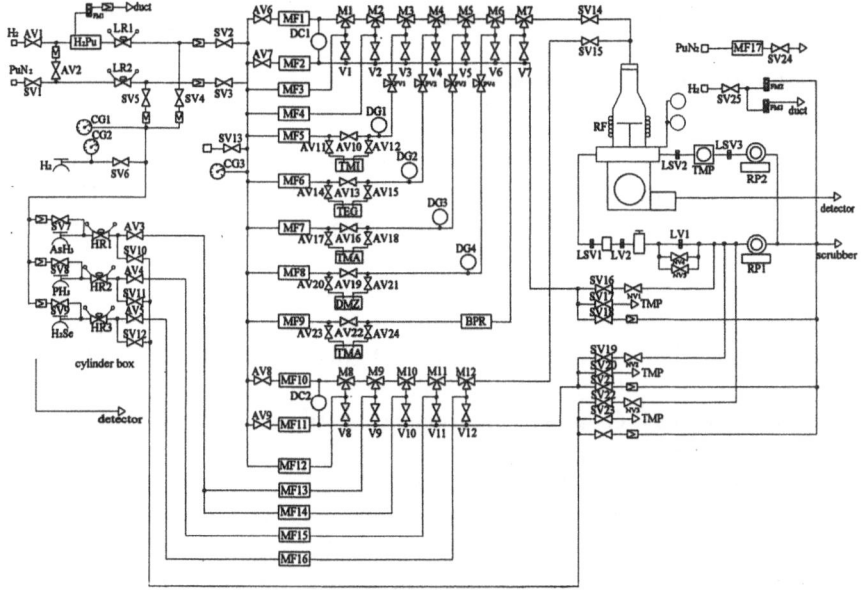

Fig.6.1. Schematic diagram of a GaAlAs/GaAs MOCVD system

ductively by Radio Frequency (RF) coils. The growth temperature is monitored by a thermo-couple inserted into the graphite susceptor. A rotary
pump and the turbo-molecular pump are used to evacuate the reactor
chamber before the growth. In order to eliminate oxygen and water contamination, a load-lock system to prevent the reactor from being exposed to the
air has been introduced.

Fig.6.2. Photograph of a MOCVD system

6.1.2 Example of MOCVD Growth

a) A Double-Heterostructure Wafer

Substrates were prepared from a Si-doped GaAs wafer with $\langle 100 \rangle$ orientation titled 2° towards the $\langle 111 \rangle$ direction, by scribing them into typically a 1 cm² area. The surfaces of the substrates were etched by a H_2SO_4: $H_2O:H_2O_2$ (3:1:1) etchant for one minute before loading it into the reactor. Five epitaxial layers having Double-Heterostructure (DH), i.e. an n-type GaAs buffer layer (Se doped 0.5 μm), n-type $Ga_{0.71}Al_{0.29}As$ cladding layer (Se doped, 2 μm), p-type GaAs active layer (Zn doped, 3 μm), p-type $Ga_{0.71}Al_{0.29}As$ cladding layer (Zn doped, 1 μm, and a p-type $Ga_{0.9}Al_{0.1}As$ cap layer (Zn doped, 0.2 μm) were successively grown on the substrate at a temperature of 780°C under atmospheric pressure. Typical growth conditions are summarized in Table 6.1.

A SEM photograph of a cleaved and etched cross section of a grown DH wafer is pictured in Fig. 6.3. The growth rate for this wafer was about $0.05 \div 0.07$ μm/min. The grown wafers were characterized by PhotoLuminescence (PL), X-ray diffraction and Hall measurements. The PL spectrum at room temperature, excited by the HeNe laser ($\lambda = 6328$Å) is comparable to that obtained from LPE wafers, as illustrated in Fig. 6.4. The alloy composition x of the $Ga_{1-x}Al_xAs$ cladding layer was determined to be 0.29 from the 0.69 μm PL peak and the following relation from Chap. 2.

$$E_g = 1.424 + 1.247x \quad [eV] . \tag{6.1}$$

The determined alloy composition agrees with the mole flow ratio of [TMA]/([TMA] + [TMG]) (see the definitions in Sect. 4.3.3) as well as the value obtained from the lattice parameter measured by X-ray diffraction.

Table 6.1. Typical growth conditions of a GaAlAs/GaAs DH wafer with a thick active layer (pressure $P_s = 760$ Torr, growth temperature $T_s = 780$°C)

		N-clad layer	Active layer	P-clad layer	Cap layer
Flux [mol/min]	TMG (-11°C)	$2.2 \cdot 10^{-5}$	$2.2 \cdot 10^{-5}$	$2.2 \cdot 10^{-5}$	$2.2 \cdot 10^{-5}$
	TMA (19°C)	$8.5 \cdot 10^{-6}$	–	$8.5 \cdot 10^{-6}$	$2.9 \cdot 10^{-6}$
	DMZn (-40°C)	–	$3 \cdot 10^{-6}$	$3 \cdot 10^{-6}$	$2.6 \cdot 10^{-5}$
	AsH₃ (20%)	$1.16 \cdot 10^{-3}$	$1.16 \cdot 10^{-3}$	$1.16 \cdot 10^{-3}$	$1.16 \cdot 10^{-3}$
	H₂Se (400ppm)	$5.4 \cdot 10^{-7}$	–	–	–
Hydrogen flux [ℓ/min]		8	8	8	8
Group-V/group-III		38	53	38	47
Growth time [min]		25	45	12.5	3

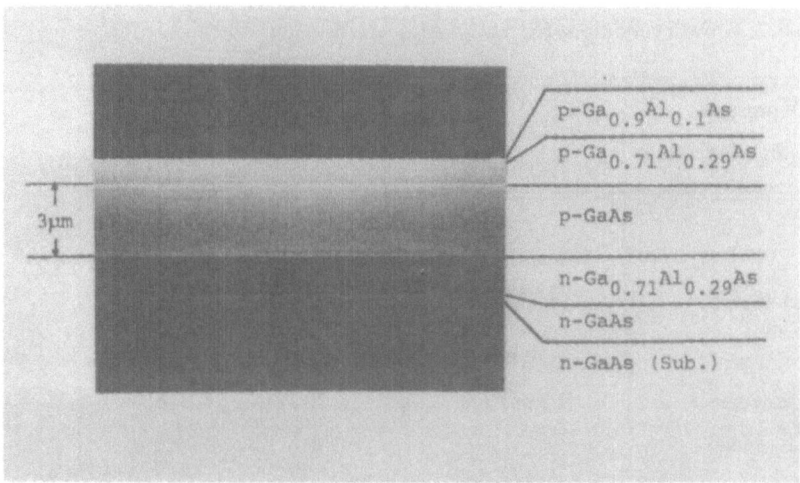

Fig.6.3. Cross-sectional SEM photograph of a GaAlAs/GaAs DH wafer with a thick active layer grown by MOCVD

The doping condition is an important parameter, influencing the threshold current of a DH wafer. Doping concentrations were determined using the Van der Pauw method: The donor concentration of the Se-doped cladding GaAlAs layer was $7 \cdot 10^{17}$ cm^{-3} under a molar flow ratio of $[H_2 Se]/[AsH_3] = 3 \cdot 10^{-4}$. In addition, the acceptor concentration of a Zn-doped cladding GaAlAs layer was estimated to be $7 \cdot 10^{17}$ cm^{-3} under a molar flow ratio of $[DMZn]/([TMG]+[TMA]) = 0.23$.

b) Semiconductor Multilayer Reflector

For the purpose of realizing a DBR Surface Emitting (SE) laser, which we believe to be one of the basic opto-electronic devices (discussed in Chap. 10), we grew a DH wafer with a 3-μm thick GaAs active layer sandwiched

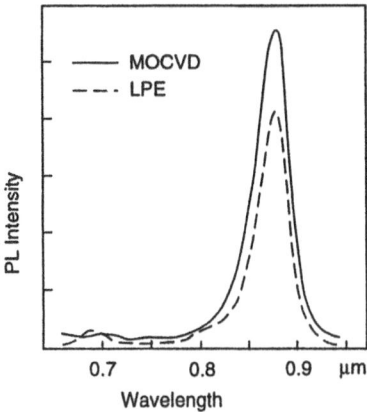

Fig.6.4. Photoluminescence spectra of a GaAlAs/GaAs DH wafer grown by MOCVD in comparison with that grown by LPE

Table 6.2. Growth conditions for a $Ga_{0.9}Al_{0.1}As/AlAs$ multilayer

			$Ga_{0.9}Al_{0.1}As$	AlAs
Growth pressure	P_g		760 torr	
Susceptor temperature	T_s		780°C	
Mole flow rate	TMG (−11°C)		$2.2 \cdot 10^{-5}$	--
[mole/min]	TMA (19°C)		$2.9 \cdot 10^{-6}$	$2.4 \cdot 10^{-5}$
	AsH_3 (20 %)		$1.16 \cdot 10^{-3}$	$1.16 \cdot 10^{-3}$
Total flow rate	[ℓ/min]		8	8
Growth time(s)			73	83

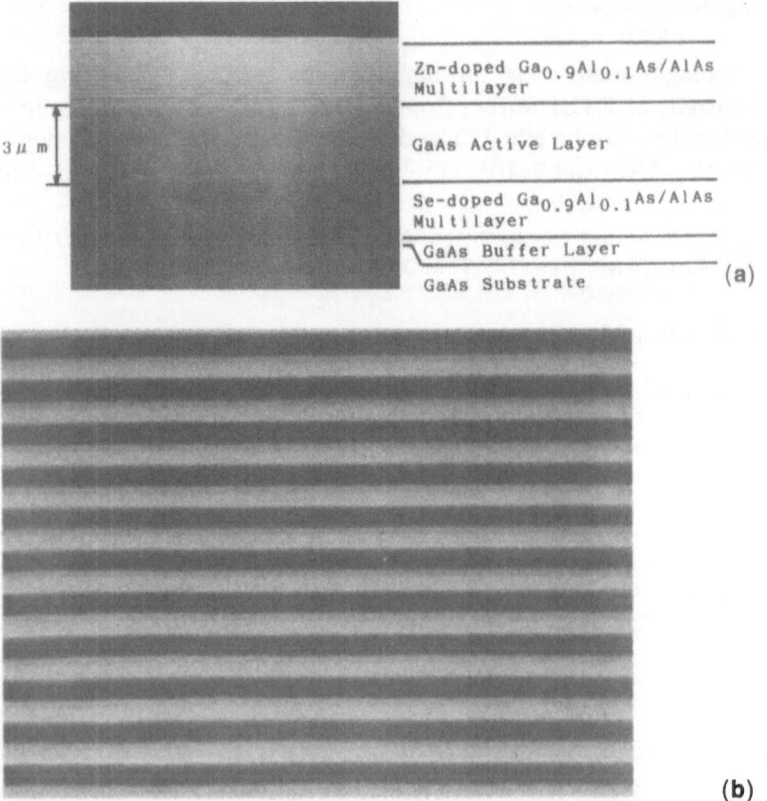

(a)

(b)

Fig.6.5. (a) Cross-sectional SEM photograph of a wafer with $Ga_{0.9}Al_{0.1}As/AlAs$ multilayer Bragg reflectors and (b) a part of the multilayer

by a couple of periodically layered Bragg reflectors composed of 30 layers of alternating GaAs and AlAs. Table 6.2 lists the growth conditions. The doping conditions are the same as those of the DH wafer, discussed in the former subsection, therefore the carrier concentration in the multilayers is estimated to be of the order of 10^{18} cm^{-3}. A SEM photograph showing a cross-sectional view of a Bragg reflector is depicted in Fig.6.5, with a 1600 Å periodicity. The uniform growth of AlAs indicates that the degradation of the crystal quality due to the oxygen contamination is negligible.

6.1.3 Characterization

a) Evaluation of the Nominal Threshold-Current Density

DH wafers grown by MOCVD were evaluated by measuring the nominal threhold-current density J_{th}/d of a 17-μm oxide stripe laser under pulsed conditions at room temperature. Typical light output-current (L-I) chracteristics of a stripe laser with a cavity length of 475 μm are illustrated in Fig.6.6. The nominal current density was calculated by dividing the threshold current by the product of the cavity length, active layer thickness and lasing spot width, as measured by near-field patterns. By changing the cavity length within the range of 45 ÷ 600 μm, the threshold-current density was measured against the mirror loss given by $\alpha_m = \ln(1/R)/L$ (R: reflectivity of the facet, L: cavity length), and is plotted in Fig.6.7. With a best-fitting procedure, the nominal threshold current density J_{th}/d [kA/(cm$^2 \cdot \mu$m)] can be expressed by the mirror loss α_m [cm^{-1}] as follows:

$$J_{th}/d = A(\alpha_m + \alpha_1)^2 \tag{6.2}$$

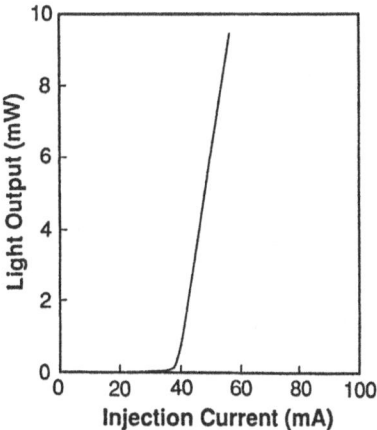

Fig.6.6. L-I characteristics of stripe GaAlAs/GaAs lasers

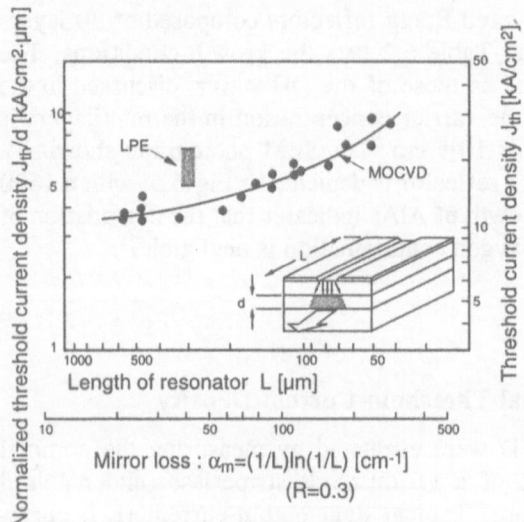

Fig.6.7. Nominal threshold current density J_{th}/d of oxide stripe lasers with a 3 μm thick active layer grown by MOCVD versus cavity length

with A = $2 \cdot 10^{-5}$ kA/μm and α_1 = 400 cm^{-1}. The minimum nominal threshold current density was 3.6 kA/(cm$^2 \cdot \mu$m), which is about 20 % lower than that obtained from material grown by LPE.

b) Reflectivity of a Multilayer Bragg Reflector

The reflectivity of a multilayer Bragg reflector composed of 30 GaAs/AlAs layers was measured on the side of the crystal surface. Figure 6.8 graphs the measured reflectivity against the wavelength. The maximum reflectivity

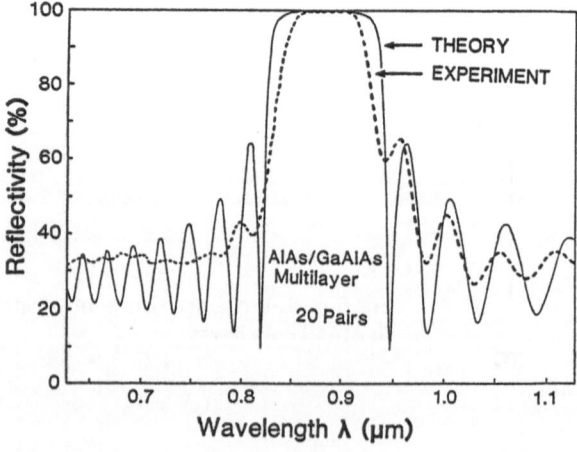

Fig.6.8. Measured spectral reflectivity of a 15 pair $Ga_{0.9}Al_{0.1}As$/AlAs multilayer

was 96%, obtained at 1.02 μm for an initial growth. Presently, a value of 99% or higher has been achieved.

6.2 Molecular-Beam and Chemical-Beam Epitaxy

6.2.1 Background

Molecular-Beam Epitaxy (MBE) and Metal-Organic Chemical-Vapor Deposition (MOCVD) are established compound semiconductor growth techniques. Although MBE and MOCVD are two important epitaxial growth techniques at present, they have advantages and disadvantages for device applications [6.4].

Chemical Beam Epitaxy (CBE) is a relatively new epitaxial technology and it incorporates both metal-organic sources for the group-III elements and hydride sources for the group-V elements. CBE combines the advantages found in the MBE and MOCVD techniques to establish new grounds. CBE enables epitaxial layers to be grown in the molecular flow region of a vacuum, but material sources can be supplied by the gaseous phase located outside the growth chamber. This feature permits versatility and convenience in exchanging source materials.

All sources used in MBE are solid, and epitaxial layers are grown in an ultra-high vacuum ($<10^{-9}$ Torr) system [6.5-9]. Flat and abrupt heterojunctions and ultra-thin multilayer structures can be expected from material grown by MBE [6.10-12]; however, "oval defects" are present on epitaxial layers [6.9, 13-15]. MBE also exhibits difficulties in growing alloy semiconductors such as GaInAsP with high-vapor-pressure materials like P and As [6.16]. Temperature control of the effusion cells in the conventional MBE system is particularly difficult, inhibiting control of the precise flux ratio of As to P throughout the growth period.

MOCVD [6.17-20] uses gaseous reactants for epitaxial growth. Metal-organic sources and hydrides are controlled by the Electric Mass Flow Controllers (EMFCs), and hence a relatively high growth rate and mute-wafer scale-up can be expected. There is a variety of sources available for a variety of applications. However, the presence of flow patterns may, some times, provoke poor uniformity in alloy compositions since the pressure inside a reactor becomes higher than 10^{-2} Torr. The use of gaseous dopants may cause some "memory effect".

Because the CBE technique uses cracked hydride gases for the group-V materials instead of uncracked gases as in MOCVD, or thermally decomposed solid sources as in MBE, P_2 and As_2 dimers are the main elements for epitaxial growth. In MBE systems, the group-V molecular beams usually consist of P_4 and As_4 [6.21], but P_2 and As_2 are said to be better for growing high-quality crystals. In the MOCVD process, hydrides and alkyls are thermally decomposed above the substrate surface so that there is a possibility of the adduct product formation.

CBE can take advantage of the molecular-beam nature for high thickness and composition control [6.22-24], since typical pressures in a CBE growth chamber are less than 10^{-4} Torr. Additionally, the dominant background gas H_2 is beneficial for the growth of high-quality crystals [6.25]. The amount of these hydride gases can easily be controlled by EMFCs, like in the MOCVD system. The growth process of the CBE technique is not limited by the diffusion of the group-III alkyls as in the MOCVD process. Rather, it is limited by the alkyls arrival rate, and hence a higher growth rate is achieved [6.23]. Metal-organic sources are also introduced as molecular beams into the chamber and pyrolytically decompose on the wafer surface. Therefore, the CBE growth mechanism employs no complex hydrodynamic boundary layer near the surface of the substrate holder, which exists in the MOCVD process. Neither flow patterns nor a dopant memory effect are expected. Hence, the CBE technique has merits for MBE of high-quality crystal structures and interfaces, as well as the advantages of MOCVD for controllability and versatility of sources. For particular device applications, for instance, Surface Emitting (SE) laser diodes with semiconductor Distributed Bragg Reflectors (DBR) [6.26], a relatively high growth rate, thickness uniformity, and heterojunction abruptness are required at the same time. The previous discussion indicates the suitability of CBE for such device applications.

Historically, the idea of using gas sources for MBE was originated by *Morris* and *Fukui* [6.27] of AT&T Bell Laboratories for the introduction of PH_3 and AsH_3 into the MBE system. Since then, *Panish* [6.28] of AT&T Bell Laboratories and *Calawa* [6.29] of the Massachusetts Institute of Technology used the same method to grow GaInAs and GaAs, and *Panish* called the MBE system using hydride gases for the group-V sources a Gas Source MBE (GSMBE) [6.30-33]. In Germany, *Veuhoff* [6.34] also managed to apply $Ga(CH_3)_3$ and AsH_3 in MBE for GaAs growth and called this method a MetalOrganic MBE MOMBE). The first time the name "Chemical-Beam Epitaxy" (CBE) appeared was at the Third Int'l Conf. on MolecularBeam Epitaxy by W.T. Tsang of AT&T Bell Laboratories in 1984. He grew some III-V compound semiconductors with a modified MBE system [6.24, 35-37] and demonstrated the applicability of these materials for devices [6.37-39]. In 1983, a group at the Tokyo Institute of Technology showed the possibility of selective epitaxy of GaAs with trimethyl gallium [6.39]. Although other studies of the CBE technique have been conducted by many other laboratories [6.41-44], this technique has not fully matured. For example, there is no report on the general behavior in the growth of $Ga_x In_{1-x} As_y P_{1-y}$, except by W.T. Tsang; there is not enough data for the growth of InP for different PH_3 flow rates, and there was no doping incorporation study for both InP and GaInAsP with Be and Si before 1989.

6.2.2 Chemical Beam Epitaxial System

A CBE system resembles a conventioanl MBE machine [6.45] because of their common design background, as discussed in the previous section. A difference is seen between the CBE and MBE gas cells and the gas-handling systems. Although effusion cells are used for solid sources in MBE, there are also gas cells: a cracking cell for the hydride gases and a slow temperature cell for metalorganic sources. The two types of cracking cells available today are a high-pressure type [6.26] and a low-pressure type [6.27]. However, effusion cells are still the major method of evaporating solid dopant materials. There are presently two choices of the main vacuum units: a turbo molecular pump or a diffusion pump. The former has a higher pumping efficiency and less oil contamination than the diffusion pump and the latter has a higher reliability and easier maintenance. The two ways to control the flow of the gaseous sources is either by an Electric Mass Flow Controller (EMFC), or by an electric pressure controller. Using a low-pressure cracking cell, and EMFCs for the group-V hydride gases, the high-pressure side of the cell is preadjusted to about 200 Torr. For the group-III metalorganic sources controlled by the EMFC, one can either use the vapor pressure of the source only, or combine it with H_2 carrier gas. The carrier gas is usually controlled between 20 and 100 Torr before entering the metalorganic source container, called a *bubbler*, by an electric pressure controller. Large pressure differences occur between the gas entrance of cells and the growth chamber whose background pressure during growth is less than 10^{-4} Torr. This causes a molecular beam to be "shut off" by the valve instead of "blocked out" by a shutter like in the MBE system. The same precise control of molecular flows can be realized in the CBE technique [6.40].

A two-chamber configuration of a CBE system designed in this study and manufactured by Riber Co. is schematically presented in Fig.6.9a. Metalorganic sources used in this system are TriEthylGallium (TEGa) and TriMethylIndium (TMIn), with the hydrides being arsine (AsH$_3$ 100%) and phosphine (PH$_3$ 100%). Solid beryllium (Be) and silicon (Si) are employed as a p-type dopant and a n-type dopant, respectively. In the growth chamber, a quadrupole mass spectrometer, a reflected high-energy electron diffraction unit, and a flux gauge are installed for in-situ analyzation. A turbomolecular pump and a rotary pump are employed for the growth chamber, the loading chamber, and the gas vent line. In this system, a low-pressure type gas cracking cell and a low-temperature gas cell are applied. Both gas cells are made of Pyrolytic Boron Nitride (PBN) which is used for effusion cell crucibles. A photograph of the CBE machine is depicted in Fig.6.9b. An upside-down "L" shape chamber in the middle of the picture is the growth chamber, and the loading chamber is reached via a gate valve.

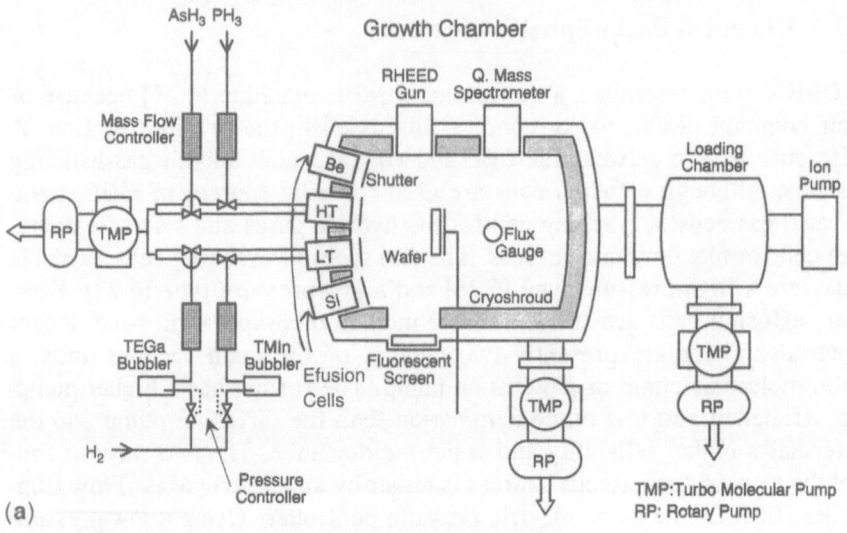

Fig.6.9. (a) A schematic diagram of a two-chamber CBE system, including gas controllers for metalorganics and hydrides. (b) A photograph of the CBE equipment manufactured by Riber Co.

6.2.3 Preparation for Growth

Temperature calibrations for the substrate are essential, and it is also necessary to repeat them after a number of growth circles. An infrared ($4\,\mu$m) pyrometer was used to measure the temperature of a molybdenum (Mo) block employed as a substrate holder and calibrated with respect to the melting point of both InSb and Al since their melting points (525°C and 660°C, respectively) are near growth temperatures in practice. The substrate temperature was monitored and corrected during growth. Readings from a thermocouple located at the back of a substrate holder are sometimes off by more than 100°C.

GaAs and InP substrates were rinsed in trichloroethylene, acetone, methanol and deionized water, and etched in a solution of $H_2SO_4:H_2O_2:H_2O$ (3:1:1), namely in case of GaAs at 65°C for 30 seconds and in that of InP at 80°C for 2 minutes. The substrate was then rinsed in deionized water for passivation [6.46] and dried by blowing nitrogen gas. In the growth chamber, the substrate was thermally cleaned to remove oxide layers on the surface above 580°C before epitaxial growth in the presence of cracked hydrides.

The vapor pressures of the metalorganics were controlled by heating the bubblers to 40°C for TM In and 35°C for TE Ga. The estimated vapor pressures at these temperatures were 3 Torr and 9 Torr for TM In and TE Ga, respectively. The pressure of H_2 carrier gas was maintained at 42 Torr for both TM In and TE Ga.

For all growths, the substrate rotation was kept at 20 rpm. The temperature of the cracing cell and the low-temperature cell were maintained at 950°C and 80°C, respectively. For material characterizations, a doping-profile plotter for the doping-level measurement, a double-crystal X-ray diffraction system for the lattice-match measurement, and HeNe- or Nd:YAG-laser excitation for the photoluminescence measurement were applied. In all growth processes we employed a valve operation to switch pases instead of a shutter operation.

6.2.4 GaAs and InP Growth

Figure 6.10 graphically depicts the GaAs growth rates versus the growth temperature for two different TE Ga and AsH_3 flows. The growth temperature was between 520°C and 600°C with a 20°C interval. Although GaAs growth rates were slowly increased from 520°C to 580°C for both cases, TE Ga was assumed to be completely decomposed in this temperature range. The growth rate increases were due to the temperature-dependent surface

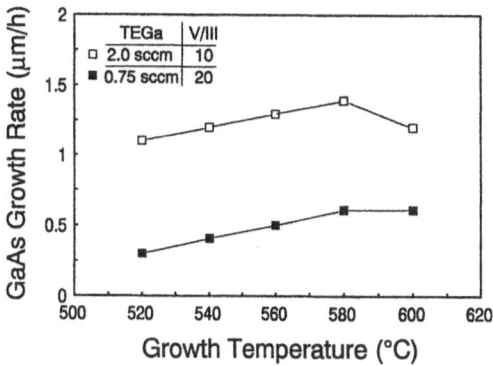

Fig.6.10. The growth rates of GaAs with two different gas flow rates plotted for several growth temperatures

diffusion of TE Ga alkyls and/or decomposed Ga. For data with a TE Ga of 0.75 sccm, the growth rate was saturated above 580°C because of the arrival rate of the TE Ga. The growth rate above 580°C for the TE Ga flow rate of 2 sccm decreased because of re-evaporation and shorter residence time of TE Ga molecules on the substrate surface. Further increases in the growth temperature would cause the desorption of the group-V elements, in addition to the re-evaporation of partially dissociated TE Ga molecules, thus causing a sharper decrease in growth rate. The lowest n-type impurity level in epitaxially grown GaAs was $5 \cdot 10^{15}$ cm^{-3}. All GaAs epitaxial layers obtained had less than $1 \cdot 10^{16}$ cm^{-3} of n-type impurity concentrations.

The behavior of InP growth rates in the same temperature range looked different from that of GaAs growth rates, although the growth mechanisms were similar, as shown ion Fig.6.11. In both cases in Fig.6.11, when growth temperatures were between 520°C to 540°C, the InP growth rate increased rapidly because of the dissociation of TM In molecules. The growth rates for the group-V to group-III ratio of 30 was saturated above 540°C because the TM In alkyl arrival rate limited the growth. For data with the group-V to group-III ratio of 19, the growth rates decreased slowly above 540°C. This decrease was explained by the re-evaporation of TM In molecules. For growth temperatures above 580°C, it is possible that there are less group-V elements so a faster decrease of the growth rate occurred. The growth rates were different for the same amount of TM In with a different amount of PH$_3$. Although this phenomenon has not been fully investigated for physical explanations, some possible causes of this phenomenon are changes in the surface diffusion length of dissociated group-III alkyls depending on the amount of both group-III and -V elements [6.47].

The InP growth rates were also examined for the different TM In flows and the substrate rotation, as plotted in Fig.6.12. A rotation of 40 rpm produced higher growth rates than that of 20 rpm. A possible explanation is an enhancement factor in the adsorption rate of the group-III alkyls for higher substrate rotation. As seen in Fig.6.12, the InP growth rates linearly increased with the increasing TM In flow rate. The highest growth rate was

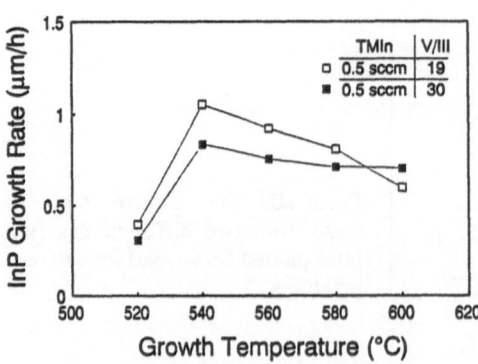

	TMIn	V/III
□	0.5 sccm	19
■	0.5 sccm	30

Fig.6.11. The growth rates of InP with two different group-V to group-III ratios plotted for several growth temperatures

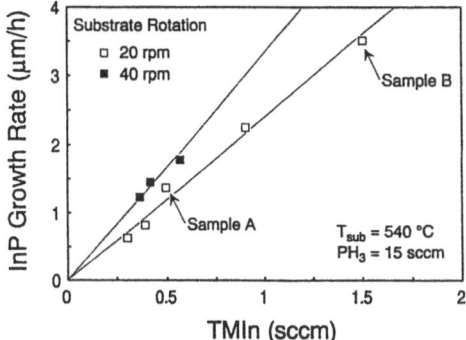

Fig.6.12. The growth rates of InP plotted as a function of TMIn flow rates for two different substrate rotations. The highest growth rate for the PH_3 flow rate was 3.5 μm/h. Photoluminescence spectra of samples A and B are shown in Fig.6.13

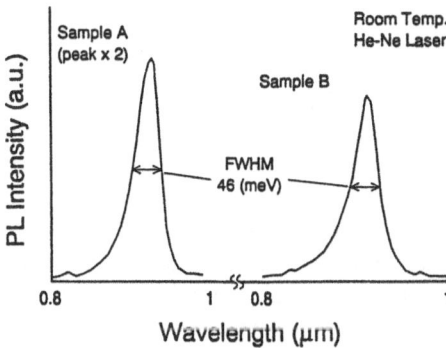

Fig.6.13. Photoluminescence spectra of InP at room temperature. Samples A and B correspond to those in Fig.6.12

3.5 μm/h and, in general, the CBE technique has a higher growth rate than that obtained by the conventional MOCVD systems [6.23].

Samples A and B denoted in Fig.6.12 correspond to the PhotoLuminescence (PL) spectra shown in Fig.6.13. Room temperature PL utilizing a HeNe laser at 6328 Å as the excitation source revealed that sample A was twice as intense as sample B. For sample B, the group-V to group-III ratio was consequently less, and not enough group-V elements were present for the growth mechanism, thus defects and/or degradations of the crystal structure caused the PL intensity to decrease. It is known that in MBE, a lack of group-V elements causes surface degradations and oval defects [6.21]. With the 15 sccm of PH_3, a good surface morphology was observed up to a growth rate of 2.13 μm/h. For both samples, the PL linewidth (full width at half maximum, FWHM) was 46 meV, the n-type impurity concentration was $1 \cdot 10^{16}$ cm^{-3}, and the peak intensity was at 0.92 μm. The lowest n-type impurity concentration obtained in InP epitaxial layers was $5 \cdot 10^{15}$ cm^{-3}.

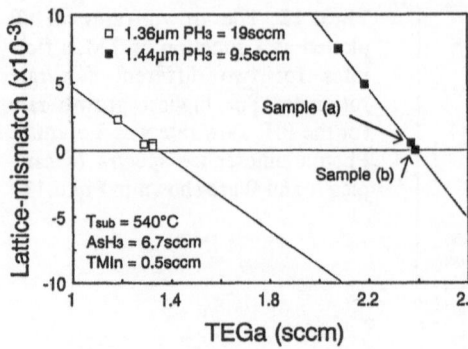

Fig.6.14. The lattice mismatch $\Delta a/a$ of $Ga_x In_{1-x} As_y P_{1-y}$ plotted as a function of TE Ga flow rates. The double-crystal X-ray diffraction spectra of the samples A and B shown in Fig. 6.15

6.2.5 $Ga_x In_{1-x} As_y P_{1-y}$ Growth

By the results obtained from the InP and GaAs growth, a growth temperature of 540°C was used for alloy compounds of $Ga_x In_{1-x} As_y P_{1-y}$. About 0.3 μm of an InP layer was first grown on a semi-insulating (Fe doped) InP substrate and then the quaternary layer of ≈ 1 μm was grown. For the lattice-match study, the flow rates of TM In and AsH_3 were fixed and the TE Ga flow rate was adjusted to obtain a lattice-match. The PH_3 flow was varied for different composition. Figure 6.14 exhibits the lattice-mismatch against the TE Ga flow rates for two alloy compositions. At the lattice-match points, one set of data was for a wavelength of 1.44 μm (x=0.34, y=0.77) which has been plotted at the left, and another was for a wavelength of 1.36 μm (x=0.3, y=0.69) which was plotted at the right. Figure 6.14 clearly shows that the lattice-match using the CBE technique is very predictable and adjustable since the behavior of the lattice-mismatch linearly depends on the TE Ga flow rates. The results were also reproducible within the range of a lattice mismatch of $\Delta a/a \leq 1 \cdot 10^{-3}$. Growth rates of those quaternary materials were about 2 μm/h and slightly different for different alloy compositions. The compositions were calculated from the wavelength of the PL spectral peak intensities.

Figure 6.15 shows the PL spectra of lattice-matched ($|\Delta a/a| < 3 \times 10^{-4}$) 1.36 μm and 1.44 μm samples. A LPE-grown 1.3 μm (x=0.27, y= 0.6) sample was measured for comparison. All samples had an impurity concentration of $1 \cdot 10^{16}$ cm^{-3} and the PL intensities were normalized. The LPE sample had the narrowest linewidth (FWHM) of 48 meV. The 1.36 μm and the 1.44 μm GaInAsP samples exhibited linewidths of 49 meV and 53 meV, respectively, which are better than those routinely available from MOCVD-grown quaternaries [6.9]. Double-crystal X-ray diffraction spectra for the 1.44 μm GaInAsP with different amounts of lattice-mismatch are depicted in Fig.6.16. Samples (a) and (b) correspond to those in Fig.6.14. Sample (a) had a mismatch of $2.3 \cdot 10^{-4}$. When the lattice was matched within the resolution limit of the measurement system (the horizontal axis is 2θ), the peaks

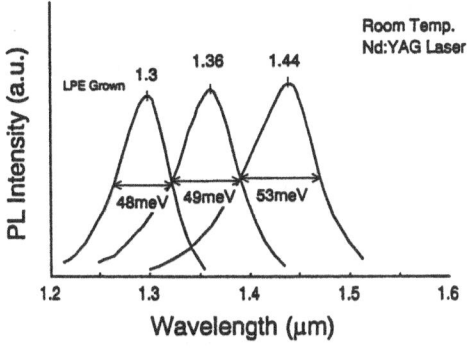

Fig.6.15. Normalized photoluminescence spectra of $Ga_xIn_{1-x}As_yP_{1-y}$ at room temperature, LPE grown $Ga_{0.27}In_{0.73}As_{0.6}P_{0.4}$ (1.3μm) is displayed as a reference

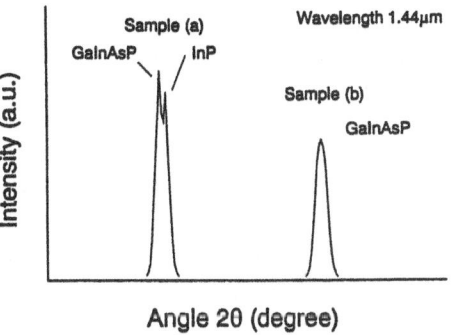

Fig.6.16. The double-crystal X-ray diffraction spectra of $Ga_{0.34}In_{0.66}$ $\cdot As_{0.77}P_{0.23}$ (1.44μm) having different amounts of lattice-mismatch. The amounts (Δa/a) are $2.3 \cdot 10^{-4}$ for sample A and within the resolution limit of the system for sample B

of the quaternary and binary coincided with each other, as seen in the spectrum of sample (b).

Ternary semiconductors of $Ga_{0.47}In_{0.53}As$ were also grown by the CBE technique and lattice-matched to InP with the mismatch being less than $1 \cdot 10^{-4}$. The lattice-mismatch values are plotted in Fig.6.17 for different TMIn flow rates and a fixed AsH_3 flow rate. Although the TMIn flow rates were different, the graphs had nearly the same slope. It is obvious that for a large TMIn flow rate, a larger TEGa flow rate is required to have a lattice-matching condition. A room-temperature PL using a Nd:YAG excitation

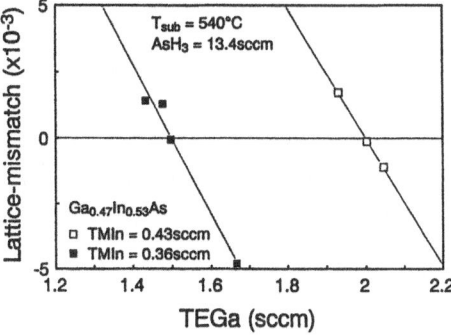

Fig.6.17. The lattice mismatch Δa/a of $Ga_{0.47}In_{0.53}As$ for different TMIn flow rates plotted as a function of TEGa flow rate

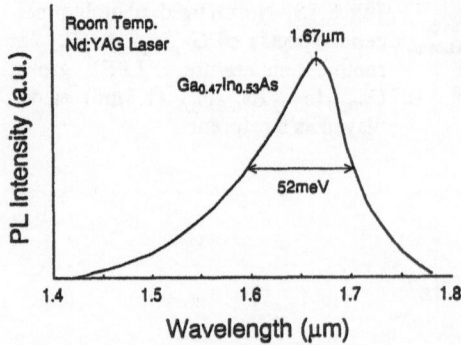

Fig.6.18. Photoluminescence spectrum of a lattice-matched GaInAs at room temperature. The impurity concentration of the sample is $9 \cdot 10^{14}$ cm^{-3}

source of a lattice-matched GaInAs epitaxial layer was measured and is presented in Fig.6.18. The peak-intensity wavelength is 1.67 nm and the spectral linewidth (FWHM) is 52 meV, which is narrower than the 75 meV obtained from a MOCVD-grown GaInAs sample [6.9]. It is known that the linewidths of ternary compounds are usually narrower than those of quaternary compounds [6.9, 24], which is the case in the present study as well. The lowest impurity level, $9 \cdot 10^{14}$ cm^{-3} (n-type), in GaInAs was achieved. Background carrier concentrations of a GaInAs layer grown by various epitaxial techniques are $3 \cdot 10^{15}$ cm^{-3} by MBE [6.48], $8 \cdot 10^{15}$ cm^{-3} by MOCVD [6.9] and $5 \cdot 10^{14}$ cm^{-3} by CBE [6.49].

6.2.6 Doping-Level Control

By utilizing effusion cells in the CBE system, well-studied dopants such as Be and Si can easily be incorporated into epitaxial growth. The advantages of using Be are that it is a well-behaved acceptor producing a shallow level above the valence band, and it can be doped at a relatively high level (in the order of 10^{19} cm^{-3}). The most attractive feature of Si as a shallow donor is its nonamphoteric behavior and very low diffusivity. We examined the in-

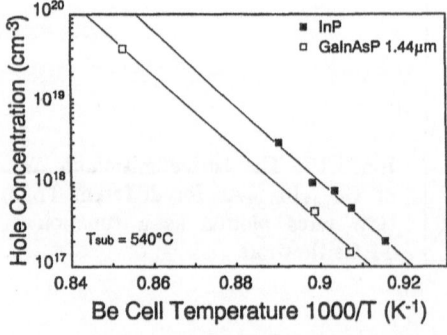

Fig.6.19. The hole concentration in InP and GaInAsP (1.44 μm) versus the inverse of Be effusion cell temperatures

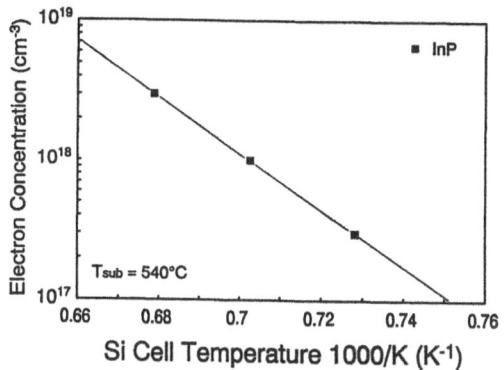

Fig.6.20. The electron concentration in InP versus the inverse of the Si effusion cell temperatures

corporation characteristics of these dopants in binary InP and in quaternary GaInAsP layers. In Figs.6.19,20, the carrier concentrations are plotted logarithmically against the inverse of the effusion-cell temperature. Figure 6.19 displays the p-type doping with Be in InP and GaInAsP ($1.44\,\mu m$). The growth rate of InP used was $1.4\ \mu m/h$ and that of GaInAsP was $2\ \mu m/h$. The level of doping might vary with the growth rates, and consequently, gas flow rates. A linear behavior was observed for both cases and Be doping levels were easily controllable as well as reproducible. A hole concentration of $3\cdot10^{19}$ cm^{-3} was achieved for GaInAsP without any degradation of the structures, a characteristic very useful for device processing. Si was utilized as a n-type dopant and the impurity concentration of Si is plotted in Fig.6.20 in the same manner as in Fig.6.19. The electron concentrations behaved linearly with Si effusion cell temperatures at around 1400 °C.

6.2.7 Summary of CBE

III-V compounds of GaAs, InP, GaInAs and GaInAsP, were grown by the CBE technique for a preliminary study of the CBE growth process. The results are summarized as follows [6.50-53]:

(i) An InP impurity level of $5\cdot10^{15}$ cm^{-3} and a GaInAsP impurity level of $1\cdot10^{16}$ cm^{-3} are very acceptable for semiconductor lasers. Further improvements can be made in reducing the concentration impurities and hence crystal Multiple Quantum Well (MQW) structures.

(ii) The growth rate of InP is a linear function of the TMIn flow rate if enough group-V elements are present and high enough growth temperatures are used.

(iii) Doping levels of 10^{17} cm^{-3} to the high 10^{19} cm^{-3} are necessary for many device structures, such as semiconductor lasers.

(iv) The lattice-match condition can be obtained using gas-flow rates, as was described in this chapter. These fundamental results prove that CBE posesses the feataure to become a major method for growing compound semiconductors, and it is especially useful for particular device applications.

(v) High-quality wafers for 1.55 μm wavelength semiconductor lasers are available, providing a value of $J_{th}/d \simeq 5$ kA/(cm$^2 \cdot \mu$m).

(vi) Low-threshold surface-emitting lasers have been obtained by utilizing the CBE techniques.

7. Characterization of Laser Materials

The preceeding chapters have described crystal growth for semiconductor lasers, with an emphasis on liquid-phase epitaxial growth techniques. Now we turn our attention to the evaluation of grown wafers and device characterization. The essential characterization of each wafer includes measuring the thickness and the band-gap energy of the epitaxial layers, observing the surface morphology, evaluating the amount of lattice-mismatch and other attributes described below.

7.1 Evaluation of Laser Wafers

In this section we describe some of the evaluation methods for grown semiconductor wafers. After the growth process, the surface morphology of wafers is routinely examined by using an optical microscope or a Nomarsky-type differential interference microscope. Mirror-like surfaces are essential for laser crystals. Figure 7.1 shows the surface of a typical GaAlAs/GaAs DH wafer. Nondistorted cross stripes reflected on the surface indicate a high-quality morphology. The size of this experimental wafer is $\approx 20 \times 13$ mm^2, but larger sizes are used in production.

A Scanning Electron Microscope (SEM) is employed to view cleaved or diagonally-ground cross sections of wafers that are stain-etched to increaase

Fig.7.1. Surface morphology of a GaAlAs/GaAs DH wafer

Table 7.1. Stain-etch solutions

GaAlAs/GaAs	NH_4OH (1 %) + H_2O_2 (30 %) = 1 : 20 , 60 s at 20°C
	$KOH + K_3Fe(CN)_6$, $\approx$ 10 s at 20°C
InP/GaInAsP	$HF + H_2O_2$, $\approx$ 10 s at 20°C
	$KOH + K_3Fe(CN)_6$, $\approx$ 10 s at 20°C

the contrast of the heterostructures. Magnifications of 1:200 000 are available, enabling recognition of an active layer with a film thickness between 200 Å to 0.5 μm. Stain-etch solutions for two different heterostructure systems are summarized in Table 7.1. Figure 7.2 exhibits a cross-sectional view of a GaAlAs/GaAs DH wafer prepared for a surface emitting laser, where an active, 3-μm thick layer is sandwiched between a 1-μm thick p-type cladding layer and a 4.5-μm thick n-type cladding layer. The SEM photograph reveals that each layer is flat and uniform.

For visual observation using higher magnifications, a Transmission Electron Microscope (TEM) was employed. Figure 7.3 shows an example of a TEM photograph taken at the boundary of a LPE-grown InP and GaInAsP system.

The compositions of grown multi-layers can be determined from the lattice constant via calibrated X-ray diffraction and from the band-gap energy, as measured by photoluminescence. Details are discussed in Sect. 7.4.

The band-gap energy E_g is determined by PhotoLuminescence (PL), as described in Sect. 7.4, or by the band edge found by measuring the absorption coefficient α of the materials.

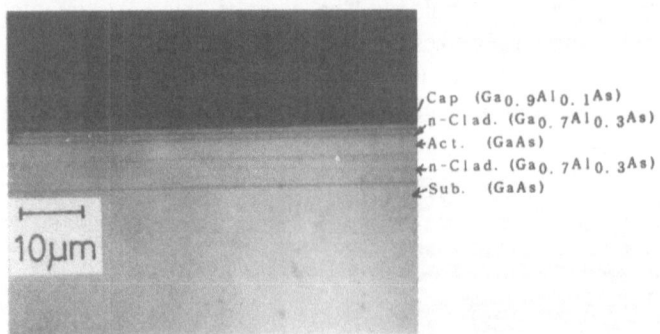

Cap $(Ga_{0.9}Al_{0.1}As)$
n-Clad. $(Ga_{0.7}Al_{0.3}As)$
Act. (GaAs)
n-Clad. $(Ga_{0.7}Al_{0.3}As)$
Sub. (GaAs)

Fig. 7.2. Cross-sectional view of a GaAlAs/GaAs DH wafer

(100)

GaInAs
2 Mono-Layers

InP
86 Mono-Layers

GaInAs
2 Mono-Layers

Fig.7.3. GaInAsP/InP DH structure and its lattice pattern by TEM

7.2 Measurement of Lattice Mismatch

The inspection of grown wafers is particularly necessary for the non-mis-matched GaInAsP system. For the measurement of lattice constants, an X-ray diffraction technique is commonly employed.

For example, consider X-ray analysis for identifying the Al content in the GaAlAs/GaAs system. There are several X-ray sources, as listed in Table 7.2, with so-called Kα and Kβ lines [7.1]. The X-ray wavelength at 1.540562 Å from Kα_1 of Cu is employed for our measurement.

Table 7.2. X-ray sources [7.1]

Material of anode	Wavelength [Å]				Accelerating volatege for K ray [kV]
	Kα_1	Kα_2	Kα (av.)	Kβ_1	
Cr	2.2897	2.2936	2.2910	2.0849	6.0
Fe	1.9360	1.9400	1.9374	1.7566	7.1
Co	1.7890	1.7929	1.7903	1.6208	7.7
Cu	1.5406	1.5444	1.5418	1.3922	9.0
Mo	0.7093	0.7136	0.7107	0.6323	20.0
Ag	0.5594	0.5638	0.5609	0.4971	25.5
W	0.2090	0.2138	0.2106	0.1844	69.5

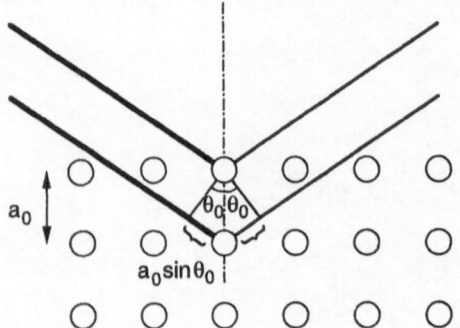

Fig.7.4. Bragg condition

a_0

$a_0 \sin\theta_0$

From Bragg's law the relation between the lattice constant a_0 and the incident angle θ_0 in terms of the X-ray wavelength λ can be expressed as (Fig. 7.4)

$$2a_0 \sin\theta_0 = m\lambda \quad \text{(m being an integer)} . \tag{7.1}$$

The lattice constant of GaAs a_0 is already known to be 5.65325 Å, so that m and θ_0 are determined to be m = 4 and θ_0 = 33.026 degrees, respectively. If the lattice constant is unknown and the reflection angle is denoted by θ_s, we have for a_s with the relation $a_0 \sin\theta_0 = a_s \sin\theta_s$

$$a_s = \frac{a_0 \sin\theta_0}{\sin(\theta_0 - \Delta\theta_s)} \tag{7.2}$$

where θ_s is equal to $\theta_0 - \Delta\theta_s$.

However, note that the lattice constant GaAlAs is larger than that of GaAs by 0.29%. Thus, the lattice constant of a GaAlAs thin film on a GaAs substrate is stretched normal to the surface, because the lattice constant of the GaAlAs layer in the horizontal direction is shrunk in order to be equal to that of the GaAs substrate (Fig. 7.5). The relationship of the GaAlAs lattice constant between the bulk crystal and the thin epitaxial layer on the GaAs substrate in the $\langle 1,0,0 \rangle$ direction is then expressed as [7.2]

$$\frac{a_r - a_0}{a_r} = 0.525 \frac{a_s - a_0}{a_s} \tag{7.3}$$

where a_r and a_s are the lattice constants of the bulk crystal and the thin epitaxial layer, respectively. The lattice constant a_r of $Ga_{1-x}Al_xAs$ can be expressed using those of AlAs (a_{AlAs}) and GaAs (a_0) as

$$a_r = xa_{AlAs} + (1-x)a_0 , \tag{7.4}$$

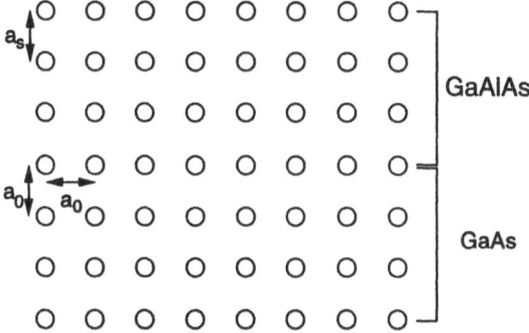

Fig.7.5. Hetero-interface of a GaAlAs/GaAs system

as explained in Chap.2. The values for a_{AlAs} and a_0 are 5.6605 Å and 5.65325 Å, respectively. Consequently, the Al content x is determined from (7.2-4). Figure 7.6 displays an example of a X-ray diffraction spectrum for a GaAlAs/GaAs DH wafer. The strongest peak is the diffraction from the GaAs substrate. The other three peaks correspond to cap, GaAlAs p-type cladding and n-type cladding layers.

7.3 Measurement of the Impurity Concentration

Several effective methods for measuring the impurity concentration in epi-taxial layers have been developed, principally in the electronic devices field. Let us introduce several practical methods:

7.3.1 Four-Point Probe Method

The four-point probe method [7.3] requires that an unknown epitaxial layer should be grown on a semi-insulating wafer, schematically shown in Fig. 7.7. The four probes should be positioned away from the edge of the wafer. A current flows into the epitaxial layer through the outer two probes, while an induced voltage is to be measured across the inner two probes. Each probe should be pressed onto the epitaxial layer with moderate pressure. If the probe space s [cm], current I [A], measured voltrage V [V] and the thick-ness of the epitaxial layer w [cm] are assumed, the resistivity ρ [$\Omega \cdot$cm] is given by

$$\rho = 2\pi s \frac{V}{I} .$$ (7.5)

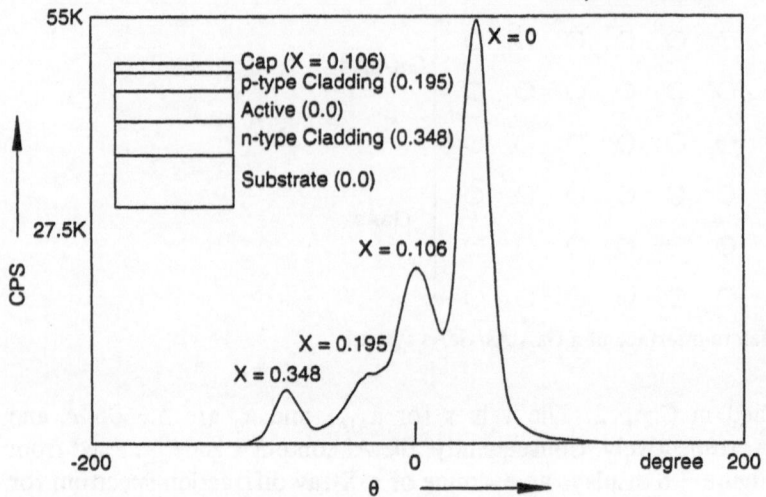

Fig.7.6. X-ray diffraction pattern of a GaAlAs/GaAs DH wafer

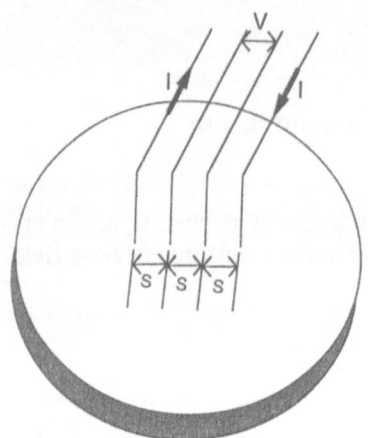

Fig.7.7. Measurement of resistivity by the four-point probe method [7.4]

This equation is valid only when w is thick compared with s. However, the typical epitaxial layer is very thin (w ≪ s), so a correction is required and (7.5) becomes

$$\rho = \frac{\pi}{\ln 2} \frac{V}{Iw} \approx \frac{4.54}{w} \frac{V}{I} . \tag{7.6}$$

The impurity concentration of the epitaxial layer can be obtained from the data for the relationship between the measured resistivity and the impurity concentration. The four-point probe method is very simple but it does not yield a precise measurement, due to the uncertainty of the measured value with the change of the probe pressure onto the wafer. Note that the four-point probe method evaluates the impurity concentration which may differ from the carrier concentration, discussed in Chap. 3. Because all impurity is not ionized. The Hall effect, described in Sect. 7.3.3, enables us to measure the carrier concentration directly [7.5].

7.3.2 Schottky Method

The basic idea behind the Schottky method is that the analysis of the capacitance/voltage characteristic of a Schottky junction gives important data such as the impurtiy concentration, the built-in potential and the width of the depletion layer. The Schottky equation for the capacitance C of a "one-sided" junction, i.e. the metal contains so many carriers that the depletion region extends only to the semiconductor, yields

$$\frac{1}{C^2} = \frac{2}{A^2 e \epsilon_s \epsilon_0 N}(V_D + V) , \tag{7.7}$$

and

$$\frac{\Delta(1/C^2)}{\Delta V} = \frac{2}{A^2 e \epsilon_s \epsilon_0 N} , \tag{7.8}$$

where A is the area of the sample, N is the carrier concentration, V_D is a built-in potential, V is a reversely applied voltage, ϵ_s and ϵ_0 are the relative dielectric constant and the dielectric constant in vacuum, respectively. The built-in potential can be determined from the intersection of (7.7) with the V lines, as shown in Fig. 7.8. From (7.8) the carrier concentration is determined from the gradient of $1/C^2$ versus voltage. Figure 7.9 plots the carrier concentration versus etching depth for a GaAs heterostructure measured by the Schottky method. Note that N is not a carrier concentration in a strict sense due to capture and emission from deep levels. But in that case meas-

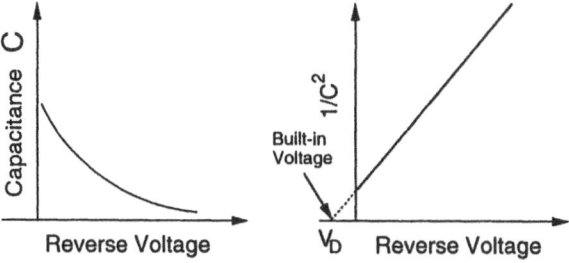

Fig. 7.8a,b. Capacitance/voltage and C^{-2} voltage characteristics of a Schottky contact

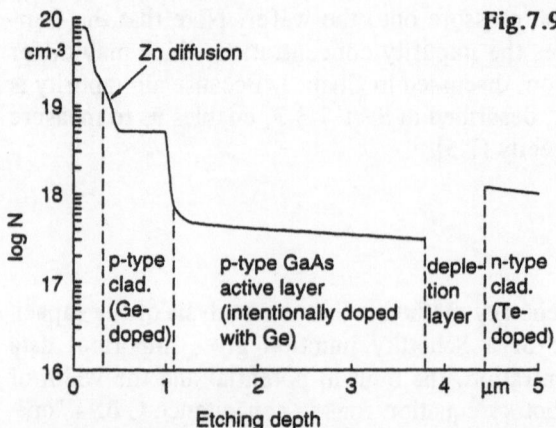

Fig.7.9. Doping profile of a DH wafer

urement at high frequency can eliminate such an effect and offers a carrier concentration at room temperature.

7.3.3 Hall Measurement

The carrier concentration can directly be determined by the Hall effect [7.6]. As an example, let us take a n-type semiconductor in which a current steadily flows. The Hall effect occurs when a magnetic field is applied perpendicular to the current flow, inducing a voltage perpendicular to both the mgnetic field and the current flow [7.5]. Figure 7.10 sketches the basic

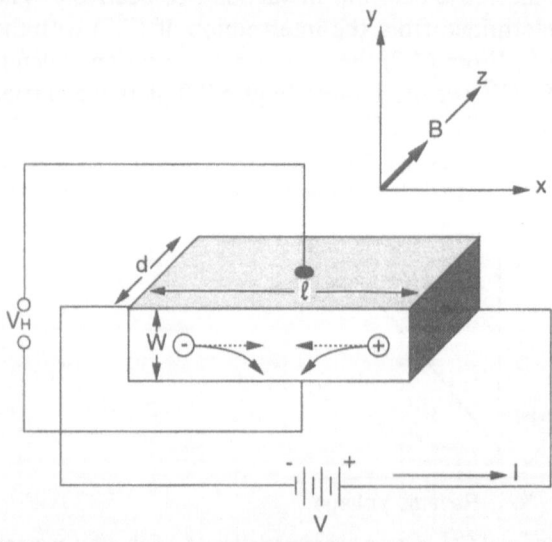

Fig.7.10. Basic experimental set-up to measure the carrier concentration using the Hall effect [7.7]

simple arrangement for measuring the Hall effect, where an electric field and a magnetic field are applied along the x-axis and the z-axis, respectively. The Lorentz force F takes place along the y axis, as depicted in Fig. 7.10 and given by (7.9)

$$F = evB \qquad (7.9)$$

where e, v and B are the electric charge, the velocity of the electrons, and the magnetic flux density, respectively. The Lorentz force moves electrons toward the $-y$ direction. Since there is no net current along the y direction, the electric field E_H, which exactly balances the Lorentz force, is along the y direction, as described in (7.10),

$$qE_H = qV_H/w = evB . \qquad (7.10)$$

Therefore, the Hall voltage is obtained as

$$V_H = vBw . \qquad (7.11)$$

The current density J_e is then given by

$$J_e = env = I/dw , \qquad (7.12)$$

where n is the carrier concentration. The Hall voltage V_H can be writen as

$$V_H = vBw = \frac{1}{en}\frac{IB}{d} = R_H\frac{IB}{d} \qquad (7.13)$$

where R_H is the Hall coefficient. The carrier concentration is obtained by

$$n = \frac{1}{eR_H} . \qquad (7.14)$$

Measuring V_H enables the carrier concentration n to be determined. The mobility μ_e [$= \sigma/(en)$] is also derived from V_H by

$$\mu_e = R_H\sigma = V_H\frac{\ell}{VBw} \qquad (7.15)$$

where the conductivity σ is

$$\sigma = \frac{I}{V}\frac{\ell}{dw} . \qquad (7.16)$$

These formula can also be applied to a p-type semiconductor; however, the polarization of V_H is reversed.

7.4 Photoluminescence

PhotoLuminescence (PL) offers distinct advantages over the more widely used variable-temperature Hall-effect measurements. First, PL can identify both the majority and minority (here, minority means of opposite type, not less of) impurities, whereaas Hall-effect measurements identify only the majority impurities and the net compensation, namely the carrier concentration. Second, PL can reveal deep impurities which may be identified by referencing published spectra, which is more difficult to achieve with Hall-effect measurements due to compensation and the uncertainty in the energy-level values yielded by the analysis. Other advantages of PL are that sample measurement and preparation are non-destructive, simpler, and more rapid than that required for Hall-effect measurements. Additionally, the thermal processing required for the application of electrical contacts is not necessary for PL samples.

The principle components of a PL experiment and a typical resulting spectrum are depicted in Fig. 7.11. By exciting the sample with above-band-gap laser light, the resulting luminescence will reveal information regarding the spatial uniformity of the emission efficiency of the grown wafers, non-

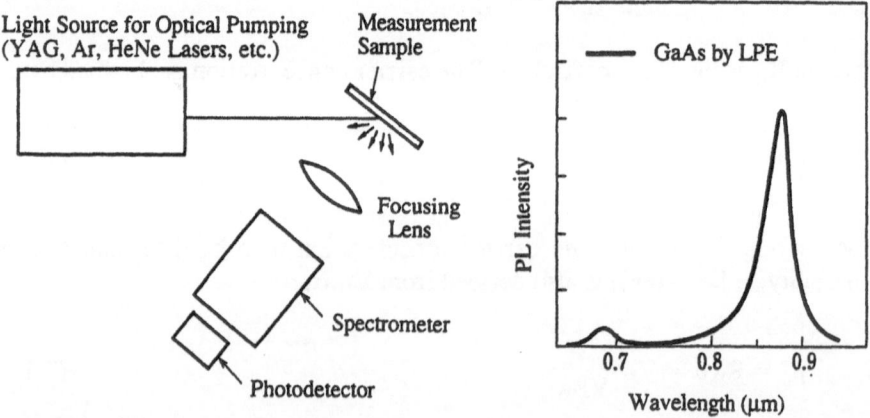

Fig. 7.11. Standard PhotoLuminescence (PL) experimental set-up and the resulting PL spectrum

emitting dark spots, and the band gaps. The PL line-widths, peak positions, and the presence of multiple peaks give important information regarding the quality of the hetero-interfaces. Additionally, the presence of growth island structure at the interfaces can be detected from splittings in the exciton PL peaks.

7.5 Measurement of the Refractive Index

Since the refractive index of a material differs according to the composition of the hetero-crystals, the values of the refractive index versus the composition are calculated from the reflection coefficient of light. A Mach-Zehnder interferometer is utilized for more precise measurements.

7.6 Misfit Dislocation

The Etch Pit Density (EPD) of the grown wafers is measured for the purpose of estimating dislocation density which affects the laser lifetime. For long-life GaAs/GaAlAs lasers, an EPD of $10^3/cm^2$ or less is required, while substrates with an EPD of 10^3 to $10^4/cm^2$ are still usable for a GaInAsP/InP system. A substrate containing a low dislocation density is desirable for obtaining high-reliability lasers.

8. Semiconductor-Laser Devices –
Fabrication and Characteristics

In this chapter, we examine the fabrication of actual laser devices and the relevant characterization techniques. The design criteria stated in Chap. 3 should be checked as the first step of device evaluation. Measurement of fundamental laser parameters such as voltage/current curves, threshold current density and spectral characteristics describe the important material properties of grown wafers. The measurement results also provide useful feedback for subsequent epitaxial growths.

8.1 Fabrication of Fundamental Laser Devices

8.1.1 Broad Contact Lasers

The first step of laser fabrication is to evaluate the quality of a wafer after it has been processed into a semiconductor laser. Let us discuss the fabrication process and the threshold-current density of a laser with a broad electrode structure, called a *broad contact laser*. Figure 8.1 illustrates the laser fabrication process, which can be categorized into the following three steps:

(i) Wafer growth.
(ii) Waveguide and resonator formation.
(iii) Electrode formation process.

For example, consider GaAlAs/GaAs materials following the growth of a heterostructure on a n-type GaAs substrate. First, the substrate is mechanically polished until the thickness decreaases to between 70 to 100 μm in preparation for cleaving. Next a 500-μm thick SiO_2 film is formed on the surface of the wafer. Then, the typally 50-μm wide and 300-μm pitch SiO_2 stripes are removed by photolithography and chemical etching. As is shown in Fig. 8.1, a 50-μm wide stripe electrode is applied to the surface of the wafer, to act as a broad contact. Typical contact materials are Au/Zn/Au for a p-type contact and Au-Ge/Au for a n-type contact [8.1,2]. A Au/Zn/Au multilayer structure is deposited by sequential evaporation of Au(300Å), Zn(500Å) and Au(2000Å) on the wafer. On the other hand, a 350 Å thick Au-Ge eutectic alloy (12wt.% Ge) and subsequently a Au film of 3000 Å thick are evaporated onto the n-type GaAs substrate. Both multi-

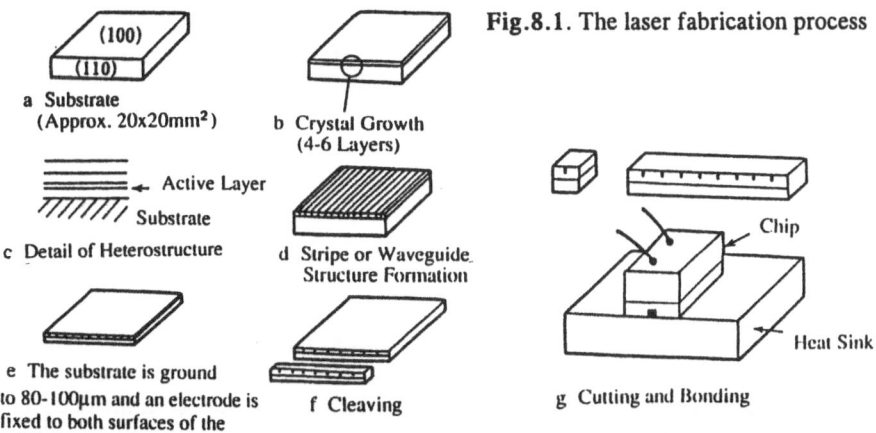

Fig.8.1. The laser fabrication process

a Substrate
(Approx. 20x20mm²)

b Crystal Growth
(4-6 Layers)

c Detail of Heterostructure

d Stripe or Waveguide
Structure Formation

e The substrate is ground
to 80-100μm and an electrode is
fixed to both surfaces of the
substrate

f Cleaving

g Cutting and Bonding

layers are alloyed at 450°C for 4 min. A contact resistance less than 10^{-5} $\Omega \cdot cm^2$ is possible when the surface of the wafer is doped over 10^{18} cm^{-3}. Following the electrode-preparation process, a laser resonator is formed by cleaving the wafer along parallel crystal planes. A razor edge is often utilized for cleaving. Occasionally, the waveguide fabrication is complete at the time of wafer growth (Chap. 9).

8.1.2 Stripe-Geometry Lasers

Usual edge-emitting semiconductor lasers have a stripe geometry to reduce the operation current and to stabilize the transverse mode. Regardless of the narrow structure, the stripe geometry features small series resistance and excellent thermal dissipation. There are various ways of making a stripe geometry and a series resistance of 0.2 to 2 Ω is obtained at an electrode with a width of $1 \div 10$ μm and a length of 250 μm.

The typical performance parameters of a GaInAsP/InP stripe laser include voltages of 0.9 V, currents of 10 to 50 mA, resistances of 1 to 2 Ω and outputs of several mW. It is common for semiconductor lasers to aim for stabilization of the transverse modes, improvement of the linearity in light output characteristics against current and also for single-longitudinal-mode operation in order to fabricate a stripe active layer with a width of $2 \div 15$ μm. Some types of fabrication methods of stripe lasers have already been established using GaAlAs/GaAs and GaInAsP/InP lasers.

As for a laser structure which aims for single-longitudinal-mode operation and its stabilization, a Distributed FeedBack (DFB) laser utilizing a diffraction grating and a Distributed Bragg Reflector (DBR) laser have been experimentally realized, and are detailed in Sect.9.2. Optical integrated circuits employing InP as a substrate are considered to be very important for future optical circuit technology. To define the proper crystal orientation of these devices, chemical etching methods and the discrimination of direction in ⟨100⟩ and ⟨011⟩, etc. are currently being studied, as shown in Sect.9.3.2.

8.2 Current Injection and Contacts

8.2.1 Current/Voltage Characteristics

When we apply a forward bias to a laser diode, we can observe its so-called *current/voltage characteristics*, which provide important information on the double-heterojunctions. The forward breakdown voltage almost corresponds to the band-gap energy of the material at which the junction is located. The differential resistance at the voltage higher than the forward breakdown represents the value of the bulk resistance and the ohmic contacts. The reverse breakdown voltage indicates the doping concentrations.

Let us discuss the current/voltage characteristics of double heterostructures [8.3] The total current injected through a p-n junction includes diffusion and drift effects. The electron and hole diffusion current per unit area, J_{Dn} and J_{Dp}, is caused by a concentration difference. However, the drift currents J_{dn} and J_{dp} are induced by the electric field at a depletion region. They are given by

$$J_n = J_{Dn} + J_{dn} = qD_n \frac{dn}{dx} + e\mu_n E(x) n , \qquad (8.1)$$

$$J_p = J_{Dp} + J_{dp} = -qD_p \frac{dp}{dx} + e\mu_p E(x) p , \qquad (8.2)$$

$$D_n = kT\mu_n/e , \quad D_p = kT\mu_p/e . \qquad (8.3,4)$$

where μ_n and μ_p are the electron and hole mobilities. At thermal equilibrium, the electronic current due to diffusion from a n-type region to a p-type region balances with the drift current by the electric field at the depletion region. Then the total electron current should be zero, as is the total hole current. If the p-n junction is forward-biased, an electron diffusion current flows into the p-type region and the hole diffusion current flows into the n-type region. Practically, many other currents are flowing due to surface effects, generation and recombination centers. However, the diffusion current is dominant at the injection level of usual light emitting devices.

Now let us calculate the diffusion current. The electron and hole concentrations, as given by an exponential approximation, were presented in Sect.3.1.3 as

$$n = N_c \exp[(F_c - E_c)/kT] \tag{8.5}$$

and

$$p = N_v \exp[(E_v - F_v)/kT] . \tag{8.6}$$

For a p-n heterojunction (p-type GaAs/n-type GaAlAs), as already depicted in Fig. 3.4, the electron minority carrier concentration n_{p0} on the p-side and the hole minority carrier concentration p_{n0} on the n-side at thermal equilibrium are given by

$$n_{p0} = (n_{i,GaAs})^2/p_0 \quad \text{and} \quad p_{n0} = (n_{i,GaAlAs})^2/n_0 . \tag{8.7, 8}$$

The intrinsic concentrations of GaAs and GaAlAs are $n_{i,GaAs}$ and $n_{i,GaAlAs}$, and the subscript 0 denotes the carrier concentrations at thermal equilibrium.

The relationship for a certain intrinsice semiconductor at thermal equilibrium, when F_c is equal to F_v, is given as

$$n = p = n_i . \tag{8.9}$$

The intrinsic carrier density n_i can be expressed as in (3.8) by

$$n_i = N_c \exp\left[\frac{F_i - E_c}{kT}\right] , \tag{8.10}$$

where F_i is the Fermi level of the intrinsic crystal. Now (3.8) can be rewritten using (8.10) as

$$n = n_i \exp\left[\frac{F_c - F_i}{kT}\right] , \tag{8.11}$$

and in a similar manner

$$p = n_i \exp\left[\frac{F_i - F_v}{kT}\right] . \tag{8.12}$$

If the voltage V_a is applied to the depletion region, we have

$$eV_a = F_c - F_v \tag{8.13}$$

Then np is obtained from (8.11-13) as

$$np = n_i^2 \exp\left[\frac{eV_a}{kT}\right] . \tag{8.14}$$

The minority carrier concentration n_p on the p-side at the edge of the depletion region (at $x = -x_p$) is therefore

$$n_p = \left(\frac{n_{i,GaAs}^2}{p}\right)\exp\left(\frac{eV_a}{kT}\right) = n_{p0}\exp\left(\frac{eV_a}{kT}\right). \tag{8.15}$$

8.2.2 Current Injection

The continuity equations of a carrier are

$$\frac{dn}{dt} = \frac{1}{e}\frac{dJ_n}{dx} - \frac{n-n_0}{\tau_n}, \quad \text{and} \quad \frac{dp}{dt} = \frac{1}{e}\frac{dJ_p}{dx} - \frac{p-p_0}{\tau_p} \tag{8.16, 17}$$

where J_n and J_p are the electron and hole current density, and τ_n and τ_p are the electron and hole lifetimes, respectively. At steady state, dn/dt and dp/dt are zero. Since there is no electric field outside of the depletion region, only the diffusion current may be considered.

$$J_n = eD_n(dn/dx), \quad \text{and} \quad J_p = -eD_p(dp/dx). \tag{8.18, 19}$$

Substituting (8.18) into (8.16) and (8.19) into (8.17), we obtain

$$\frac{d^2n}{dx^2} - \frac{n-n_0}{D_n\tau_n} = 0 \quad \text{and} \quad \frac{d^2p}{dx^2} - \frac{p-p_0}{D_p\tau_p} = 0. \tag{8.20, 21}$$

The minority-carrier concentration in the p-type region at thermal equilibrium is given by

$$n_0 = n_{p0}. \tag{8.22}$$

The solution of (8.20) is then

$$n = C_1\exp\left(-\frac{x}{(D_n\tau_n)^{1/2}}\right) + C_2\exp\left(\frac{x}{(D_n\tau_n)^{1/2}}\right) + n_{p0}. \tag{8.23}$$

From the boundary condition of $n(x=-\infty) = n_{p0}$, C_1 becomes zero, and from $n(x=-x_p)$, $n-n_p$ is obtained as

$$n - n_{p0} = n_{p0}\left[\exp\left(\frac{eV_a}{kT}\right) - 1\right]\exp\left(\frac{x_p+x}{L_n}\right) \tag{8.24}$$

where the minority carrier diffusion length L_n is

$$L_n = \sqrt{D_n \tau_n} \ . \tag{8.25}$$

The electron diffusion current density given by (8.18) is determined at $x = -x_p$ as

$$J_n = eD_n \left(\frac{dn}{dx}\right)_{x=-x_p} = -\frac{eD_n n_{p0}}{L_n}\left[\exp\left(\frac{eV_a}{kT}\right) - 1\right] , \tag{8.26}$$

and similarly, for the hole diffusion current density on the n-side

$$J_p = \frac{eD_p p_{n0}}{L_p}\left[\exp\left(\frac{eV_a}{kT}\right) - 1\right] , \tag{8.27}$$

where $L_p = \sqrt{D_p \tau_p}$. The ratio of J_n and J_p becomes

$$\left(\frac{J_n}{J_p}\right) = \frac{D_n L_p n_{p0}}{D_p L_n p_{n0}} = \frac{D_n L_p (n_{i,\text{GaAs}})^2 n_0}{D_p L_n (n_{i,\text{GaAlAs}})^2 p_0}$$

$$= \frac{D_n L_p n_0}{D_p L_n p_0} \frac{(m_{p,p} m_{n,p})^{3/2}}{(m_{p,n} m_{n,n})^{3/2}} \exp\left(\frac{E_{g,\text{GaAlAs}} - E_{g,\text{GaAs}}}{kT}\right) . \tag{8.28}$$

As $(E_{g,\text{GaAlAs}} - E_{g,\text{GaAs}})$ is usually larger than kT, electron injection ($J_n \gg J_p$) occurs. (On the contrary, if a p-type GaAlAs/n-type GaAs junction is assumed, only hole injection occurs:

For low injection conditions ($n_p \ll p_p$)

$$J = J_0\left[\exp\left(\frac{eV_a}{kT}\right) - 1\right] , \quad \text{and} \quad J_0 = \frac{eD_n n_{p0}}{L_n} . \tag{8.29, 30}$$

At high injection levels, n_p is equal to p_p at $x = -x_p$, and from (8.14)

$$n_p = n_{i,\text{GaAs}} \exp(eV_a / 2kT) \tag{8.31}$$

$$J = J_0\left[\exp\left(\frac{eV_a}{2kT}\right) - 1\right] , \quad \text{and} \quad J_0 = \frac{eD_n n_{i,\text{GaAs}}}{L_n} . \tag{8.32, 33}$$

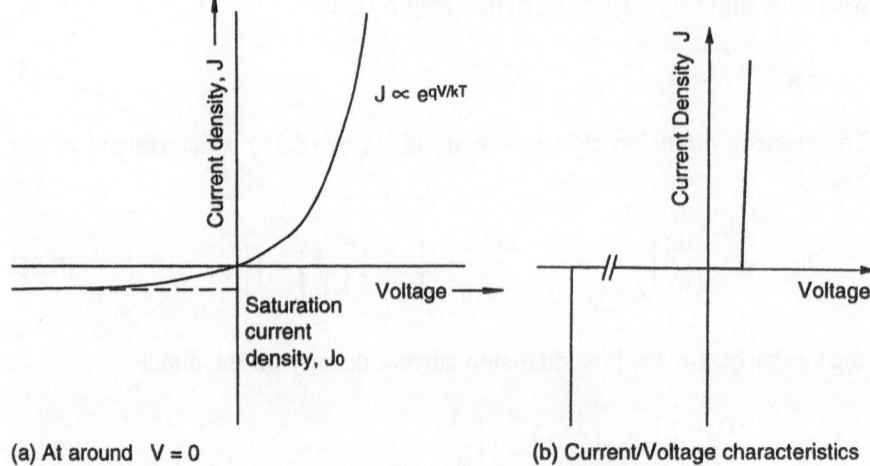

(a) At around V = 0   (b) Current/Voltage characteristics

Fig.8.2a,b. Current/voltage (I/V) characteristic of a DH laser

The current density versus an applied voltage characteristic is depicted in Fig.8.2. However, if the measurement is made using a sort of needle-type current and voltage monitors, the actual measured data is shown in Fig.8.3. The forward current density abruptly increases when the applied voltage exceeds $E_{g,s}/q$, where $E_{g,s}$ denotes the energy of the smaller band-gap heterojunction. The saturation current density is not measurable and is not seen.

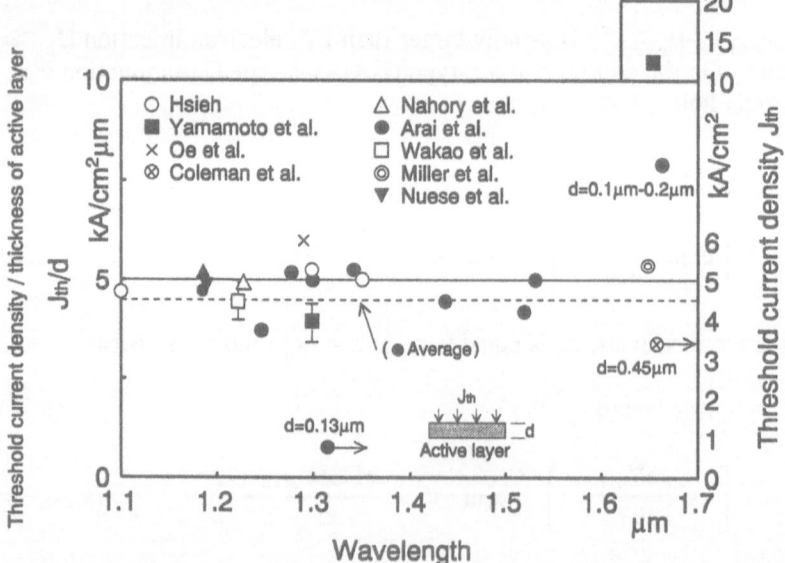

Fig.8.3. Threshold current density J_{th}/d of a $Ga_x In_{1-x} As_y P_{1-y}/InP$ laser per unit active layer width

8.3 Evaluation of the Threshold-Current Density

Measurement of the threshold current I_{th} leads to the threshold current density J_{th}, given by

$$J_{th} = \frac{I_{th}}{\ell w} \qquad (8.34)$$

where ℓ and w denote the length of a laser and the stripe width, respectively. The threshold current density J_{th} changes with a distinctive feature versus the active-layer thickness d. In other words, the minimum value exists around d = 0.1 to 0.2 μm with the minimum of J_{th} = 1 to 2 kA/cm^2. The normalized threshold current density J_{th}/d expresses the quality of the laser crystals. Figure 8.4 exhibits the threshold current density J_{th}/d of $Ga_x In_{1-x} As_y P_{1-y}$/InP per unit active layer width [8.4]. Approximately 4.5 kA/(cm$^2 \cdot \mu$m) is obtained in 1.1 < λ < 1.55 [μm] in wafers grown by liquid-phase epitaxy. Furthermore, $0.2 \div 0.5$ kA/(cm$^2 \cdot \mu$m) is obtained in GaInAs/InP grown by molecular-beam epitaxy for an emitting wavelength of 1.3 < λ < 1.55 [μm].

8.4 Gain Bandwidth and Oscillation Spectra

In semiconductor lasers, the gain spectrum has a finite width on the order of 300 Å, due to the distribution of the carriers in the bands over a range of energies $\approx$ kT, causing recombination to occur between holes and electrons having a spread of energies. Basically, the gain is determined by the density of states and the distribution function. The former expresses the number of states, and is given by a parabolic function of the wave number. The carrier distribution function is determined both by the temperature and a transient quasi-Fermi level. The optical gain of a semiconductor laser is mainly determined by the energy distribution of carriers injected into the conduction band, since the effective mass of electrons in the conduction band is smaller than that of holes in the valence band. In addition, impurity levels and relaxation times influence the optical gain. The gain width depends on the carrier density, and can be as large as 100 to 200 cm^{-1}.

In the gain-guided laser described in Table 8.1a, various modes oscillate simultaneously. When light propagates along a stripe geometry region with gain, the phase of light propagating in the gain region advances, while the phase in the lossy surroundings suffers delays. Though standing waves are formed only by orthogonal modes of plane phases at the mirrors (in

Table 8.1. Comparison between a gain-guided and an index-guided semiconductor laser

Characteristics	Gain-guided laser	Index-guided laser

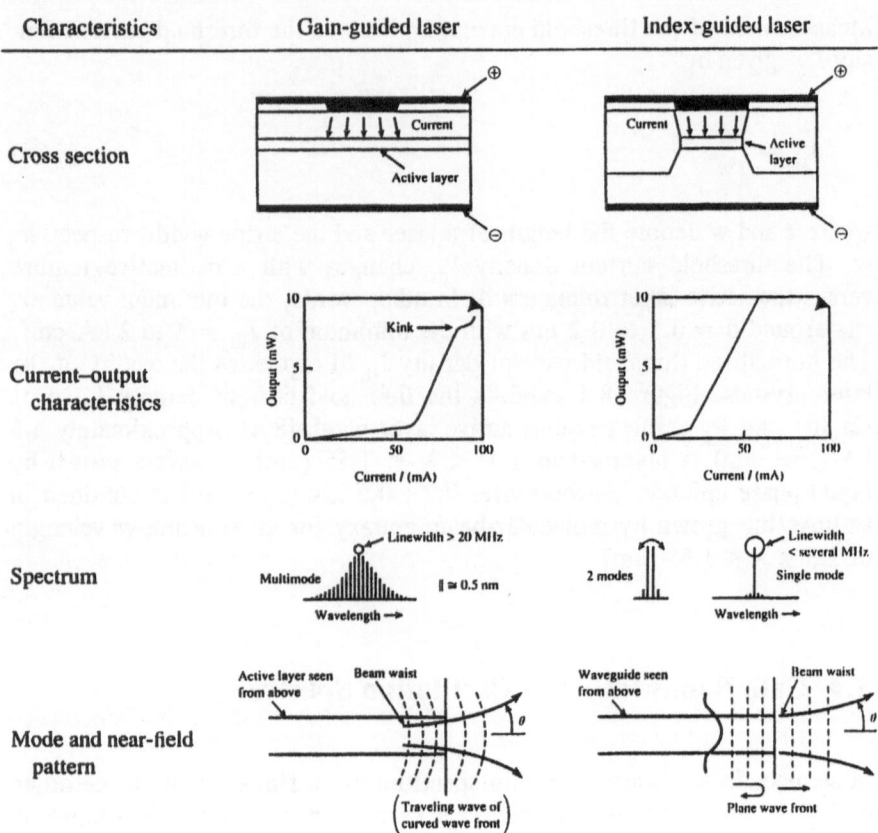

view of the result of normal modes in an index waveguide), the amplified waves discussed above are expressed only by the expansion by various modes. It is interesting to note that the total superimposed field looks like the one indicated in Table 8.1a which has a bent curvature of phase at the reflecting mirrors. Therefore, many modes can obtain gain equivalently, thus resulting in multimode oscillation and the instability of guided transverse modes at high injection levels. Consequently, a bend which is called a *kink* is sometimes observed in a light output power vs. current characteristic. However, there is little influence on the performance caused by reflecting light upon an instability in multimode-function devices. Reference is made to results by *Petermann* [8.5] in Table 8.1.

8.5 Output and Efficiency of Semiconductor Lasers

One of the most important characteristics that demonstrates semiconductor-laser performance is the light output power L versus the injected current I, which is called an *L-I characteristics*. The threshold current depends on the length of the laser resonator, reflecting power, stripe width, resonator loss, confinement factor of the gain, etc. The minimum value recorded to date is 0.55 to 2.5 mA [8.6] against a continuous-wave operation at room temperature. A buried-type index waveguide, as shown in Table 8.3b, and an excellent current confinement mechanism are necessary to obtain a low threshold, and GaAs and GaInAsP lasers with almost optimum designs can provide 10 to 30 mA.

In quantum-well lasers with an ultrathin active layer, devices with an ultralow threshold can be obtained. Whether the record of 1 mA can be broken by optimizing the structure is of world interest. A device with 0.55 mA was reported in 1987. The possibility of such a quantum-well structure is an important area of research.

As for the continuous operation of Laser Diodes (LDs), roughly speaking a few mW or more can be obtained from one facet, though there are some differences among respective kinds and types. For optical communications, the output is approximately 3 to 10 mW. Recently, outputs of 200 mW [8.7] in a single-stripe laser, 3.2 W from a wide single-strip laser and 55 W [8.8] from a 1-cm monolithic laser array (a twenty of ten-stripe structure) have been reported. The limit of the output depends on the damage of the electrodes, junction interfaces and mirror facets. The deterioration of the mirror facets is a serious problem for high output, making it necessary to apply a protective coating to the facets, which is especially necessary for GaAs lasers. Through pulsed oscillation, a peak output of several W can be obtained.

The differential quantum efficiency η_D is defined as the ratio of the increaase in the number of output photons for a given increaase in the number of injected electrons, and is written as

$$\eta_D = \frac{dP/\hbar\omega}{dI/e} \simeq \frac{dP}{dI}\frac{1}{E_g} \tag{8.35}$$

where e, E_g, $\hbar$ and ω denote the electron charge, the band-gap energy expressed in electron volts, Planck's constant divided by 2π, and the photon energy, respectively.

If the total output from both facets is P, we can obtain $\eta_D = 56\%$ for a GaAs laser by substituting $dP/dI = 0.8$ mW/mA and $E_g = 1.45$ eV into the above equation. In the case of a CW laser diode, the slope efficiency is 40 to 60 % (maximum being 75 %).

8.6 Near-Field Pattern and Far-Field Pattern

Table 8.1d displays the light intensity distribution at the edge of a laser resonator, and is called a *Near Field Pattern* (NFP). Observation is made through a microscope and an infrared monitor. It is desirable from an application viewpoint that the pattern change negligibly if the current level is changed.

The *Far-Field Pattern* (FFP) graphs the radiation characteristic of optical power from the laser. Usually it is presented with plots showing angles to the central axis, and FFPs in the horizontal direction are often different from those in vertical direction. For an ordinary stripe laser, the half angle is 15° for the horizontal direction, and 30° for the vertical direction.

The Fourier transformation of a NFP does not always correspond to that of an FFP. In the case of a gain-guided laser, the wavefront of a reflecting mirror is not always planar. In the direction vertical to an active layer, the wavefront is plane due to good waveguiding by the double heterostructure in the direction vertical to the active layer. However, in the case of the horizontal direction, since the wavefront is convex, as seen from the outside, a vertical focus seems to exist within the laser. Consequently, astigmatism occurs in the output light [8.9].

8.7 Temperature Characteristics

Semiconductor lasers are relatively sensitive to temperature. The threshold curren I_{th} and the differential quantum efficiency η_D deteriorate with increaasing temperature. The temperature reliability of the threshold current is written as

$$I_{th}(T) = I_{th}(T') \exp\left[\frac{T-T'}{T_0}\right] \tag{8.36}$$

where $I_{th}(T)$ is the threshold at T, and T_0 is called the *characteristic temperature*. T_0 = 120 to 150 K for GaAlAs/GaAs lasers and T_0 = 50 to 70 K (T' = 20°C) for GaInAsP/InP lasers. The larger the temperature characteristic is, the smaller the temperature dependence. Various causes of a small T_0, namely a poor temperature characteristic, in the GaInAsP/InP system include Auger recombination, intraband absorption, the hot-electron effects, and leakage from a hetero-barrier.

8.8 Reliability

Rapid progress in increaasing the lifetimes of GaAlAs semiconductor lasers was made within a few years before 1976. The advances were due to the development of better crystal-growth techniques, improved heat sinks, and bonding methods coupled with proper analysis of the various factors causing failure. Dark lines and dark spots in the active layer of the GaAlAs/GaAs laser appear to have a significant effect on the deterioration rate. The lifetimes which may be achieved can be estimated from accelerated tests at high tenmperatures, giving values of 10^5 to 10^6 hours. For example, in 1981, a lifetime of 10^6 was estimated after the bonding technique was improved.

On the other hand, the number of dark lines and dark spots appearing in crystals of a GaInAsP/InP laser in the 1 μm wavelength band is said to be smaller than that of a similar GaAlAs/GaAs laser, resulting in a lower deterioration rate. From the accelerated lifetime test at 70°C and a constant output of 5 mW, a lifetime of several 100000 hours has been estimated.

9. Mode-Control Techniques in Semiconductor Lasers

It is very important for semiconductor lasers to have stable transverse and longitudinal modes in order to obtain reliable performance. This is achieved by using an index waveguide structure. One of the most common techniques to control the longitudinal mode is to equip the inside of the device with a grating structure, created by overgrowth epitaxy. The LPE growth technique is the most suitable for this purpose. This is partly due to the phenomenon that LPE tends to flatten the surface, which is remarkable in comparison with other epitaxial growth techniques. This feature greatly helps in the fabrication of transverse-mode-controlled structures such as V-groove structures, buried heterostructures, Distributed FeedBack (DFB) and Distributed Bragg Reflector (DBR) structures. The theory behind stabilizing transverse and longitudinal modes and its importance will be explained. Next, structured epitaxial techniques will be summarized in Sects.9.3-6, and the fabrication of multilayer structures will be described in Sects.9.7 and 8.

9.1 Transverse-Mode Characteristics and the Single-Mode Condition

9.1.1 Necessity of Transverse-Mode Stabilization

The stabilization of the transverse mode is very effective in reducing the threshold, stabilizing the output beam and improving the modulation characteristics of a semiconductor laser. Let us start by discussing the single-mode condition. Single-mode guidance is easily achieved in the direction perpendicular to the junction because the active layer is very thin. Since the normalized index difference $\Delta = 0.008$ (Sect.9.1.3) in a typical double heterostructure, the required single-mode waveguide should be less than 0.45 μm in thickness for $\lambda = 0.8$ μm. The active-layer thickness is typically 0.2 μm and, therefore, the single-mode condition in the direction perpendicular to the junction plane is usually satisfied.

On the other hand, by forming a dielectric waveguide with an index difference in the horizontal direction and making the guiding effect larger than the gain change originating from the injected carriers, the transverse mode can be stabilized. Such lasers are called *index-guided lasers*.

Since the form of the transverse mode is determined primarily by the index difference and waveguide width, the instability of the mode due to gain changes no longer plays an important role and the power output versus the injected current linearly increases without kinks. As for spectra, these lasers operate in a single mode and are coherent. Occasionally, an instability due to reflected light from optical components in a practical system may occur. Figure 9.1 depicts the following three types of index waveguides.

(a) **Buried type**: A double heterostructure wafer grown in advance is processed into a mesa by etching. The second crystal growth produces side walls, therefore the mesa is buried with semiconductors having a small refractive index

(b) **Selective growth type**: A double heterostructure is grown onto a substrate which has been processed into a groove, terrace, or mesa in advance.

(c) **Distributed impurity type**: The refractive index difference is formed through an impurity density variation in the horizontal direction.

Most semiconductor lasers used in light-wave communication need higher coherence, and index-guided lasers are essentially utilized for improving various lasing characteristics. However, the gain of the fundamental mode decreases with the increase of output power, causing the higher-order modes to oscillate. Consequently, it is an important goal in laser design to extend the single-mode region to higher output levels.

On the other hand, optical discs require resistivity against the noise induced by reflected light, and in such cases, a multimode structure is intentionally utilized in order to decrease the mode-hopping noise.

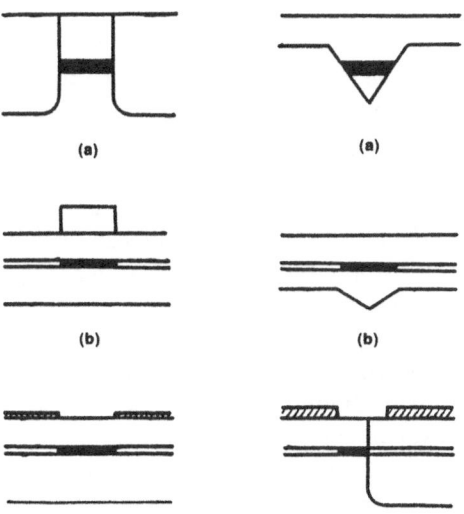

Fig.9.1. Representatives of index waveguides (a) buried type, (b) selective growth type, (c) distributed impurity type

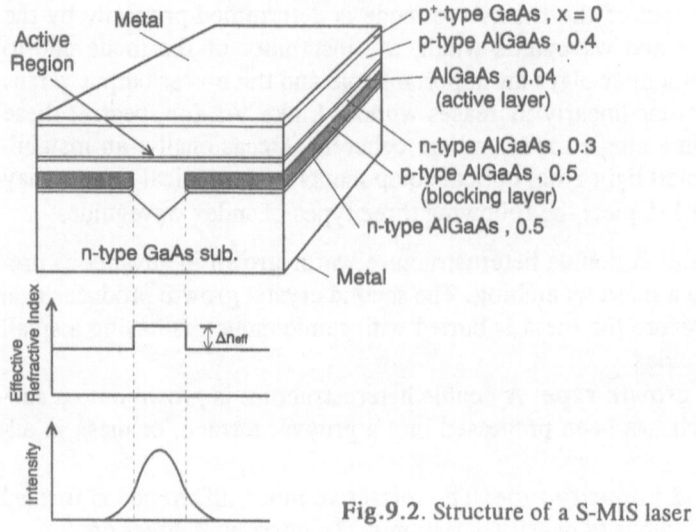

Fig.9.2. Structure of a S-MIS laser

9.1.2 Equivalent Refractive-Index Method

Let us consider the single-transverse-mode condition. The equivalent refractive-index method [9.1] greatly helps us to simplify the problem, by treating a complicated dielectric waveguide as an equivalent uniform planar waveguide. The equivalent refractive index is represented by the propagation constant β divided by wave number k_0 in the vacuum, i.e.,

$$n_{eq} = \beta/k_0 . \tag{9.1}$$

Now, let us apply the equivalent refractive-index method to index-guided structures. Figure 9.2 delineates the structure of a S-MIS laser [9.2] as an example of a transverse-mode-controlled semiconductor laser. Light is confined in the active region not only due to the waveguide structure perpendicular to the junction, but also due to a horizontal waveguide structure formed by the thickness difference of a n-type $Ga_{0.7}Al_{0.3}As$ layer between, on, and beside the V-groove. The p-type $Ga_{0.5}Al_{0.5}As$ layer blocks the current flow. Thus, the driving current flows only through the V-groove. Typical transverse-mode-controlled semiconductor lasers contain both light and current confining structures.

The waveguide structure can approximately be represented as in Fig. 9.3a. Let us divide the structure into two regions, A and B. The refractive index profile represented by the equivalent refractive index of each region is shown in Fig.9.3b. The model can be simplified to the three-layer slab waveguide (Fig.9.3c). The single-transverse-mode condition is therefore expressed as in Sect.3.1.4:

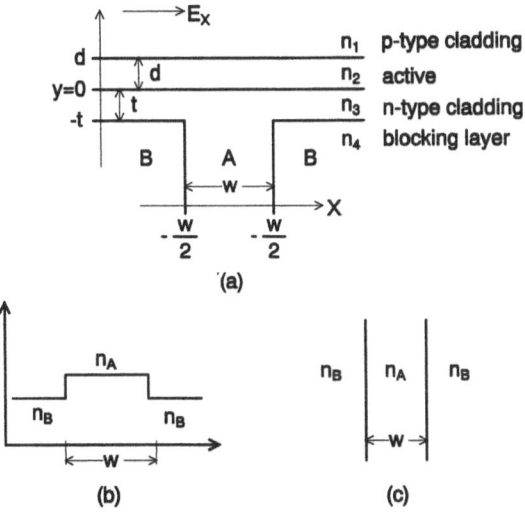

(a)

(b)   (c)

Fig.9.3a–c. Waveguide model of a S-MIS laser

$$W < \frac{\lambda_0/2}{\sqrt{n_A{}^2 - n_B{}^2}} \qquad (9.2)$$

where n_A and n_B are equivalent refractive indices in the regions A and B, respectively.

9.1.3 Eigenvalue Equation of a Guided Mode

The equivalent refractive indices of the regions A and B are determined by the propagation constants obtained from the eigenvalue equation of a guided mode. The aim of this section is to theoretically determine the propagation constants.

The electric field E_x in the region B is expressed as

$$E_x = \begin{cases} A\,e^{-p(y-d)} & (d \le y)\,, \\ B\cos(hy) + C\sin(hy) & (0 \le y \le d)\,, \\ D\,e^{qy} + E\,e^{-qy} & (-t \le y \le 0)\,, \\ F\,e^{r(y+t)} & (y \le -t)\,, \end{cases} \qquad (9.3)$$

where A, B, C, D, E and F are intensities, and p, h, q and r are the short-hands

$$p = \sqrt{\beta^2 - k_0^2 n_1^2}\ , \quad h = \sqrt{k_0^2 n_2^2 - \beta^2}\ ,$$

$$q = \sqrt{\beta^2 - k_0^2 n_3^2}\ , \quad r = \sqrt{\beta^2 - k_0^2 n_4^2}\ . \tag{9.4}$$

and d and t are the thicknesses of the active and n-type cladding layers, respectively, and w is the channel width. From the electric- and magnetic-field boundary conditions, the eigenvalue equation can be expressed as [9.3]

$$\tan(hd) = \frac{p + q + (p-q)V}{h - pq/h + (h+pq/h)V}\ , \tag{9.5}$$

$$V = \frac{q-r}{q+r}\exp(-2qt)\ . \tag{9.6}$$

The eigenvalue equation in region A can be obtained for the thickness t being assumed infinite ($t \to \infty$). Then,

$$\tan(hd) = \frac{p + q}{h - pq/h}\ . \tag{9.7}$$

The propagation constants of the regions A and B are determined from (9.5-7). The equivalent refractive indices of regions A and B can be derived from (9.1).

The single-transverse-mode condition is presented in Fig.9.4 [9.2]. Δ_{AC} and Δ_{CB} are

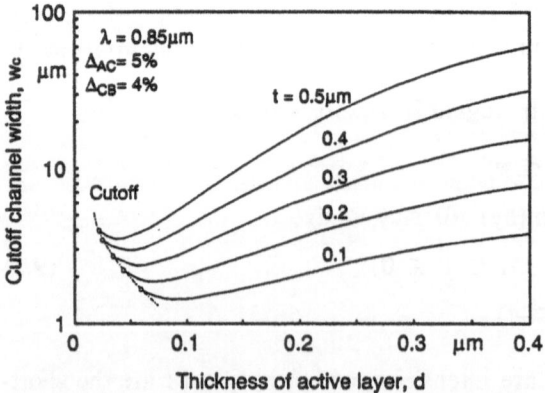

Fig.9.4. Channel width W_c of the higher-mode cuttoff condition against active layer thickness d with a cladding layer thickness t

$$\Delta_{AC} = \frac{n_2{}^2 - n_3{}^2}{2n_2{}^2}, \quad \Delta_{CB} = \frac{n_3{}^2 - n_4{}^2}{2n_3{}^2}. \tag{9.8,9}$$

For example, when it is assumed that d = 0.2 μm and t = 0.3 μm, the cutoff channel width w_c is 5.9 μm at λ = 0.85 μm. Consequently the channel width should be less than 5.9 μm in order to satisfy the fundamental mode condition. In actual devices, an injected-carrier profile along the laterial direction eases the restriction of the channel width.

9.2 Longitudinal-Mode Control

In 1978 it was found that a silica fiber exhibits ultra-low loss at 1.55 μm. Although researchers recognized the importance of the 1.55 μm band for long-distance transmission, material dispersion (3.5ps/(km·Å)) is large at that wavelength and the transmission-band width is limited. Y. Suematsu pointed out that a single longitudinal mode operating at 1.55 μm is a requirement to achieve long-haul and wide-band transmission, even when the laser is modulated.

In a conventional Fabry-Perot type laser, however, the number of long-itudinal modes increases drastically when the laser is modulated at a high frequency, even if the laser can operate with a single mode in CW (Fig. 9.5). This can be understood by the fact that there is no difference in the loss in different modes and that one mode happens to reach the threshold in CW operation because of only a small difference in gain. It is necessary to

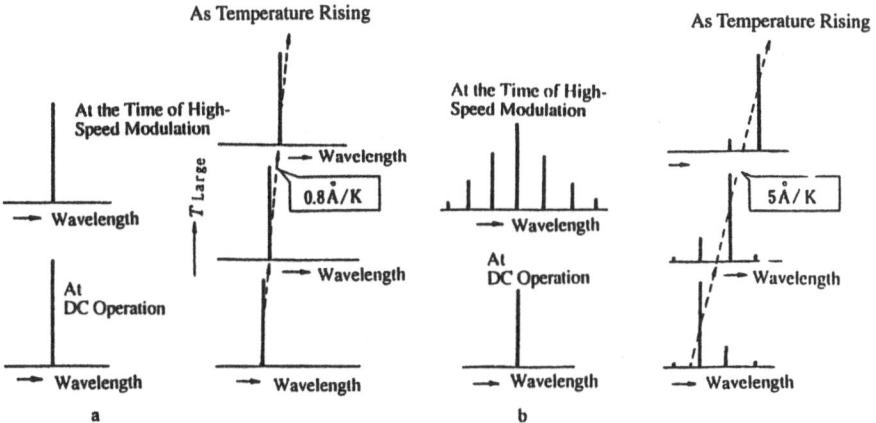

Fig.9.5a,b. Spectra of Dynamic-Single-Mode (DSM) lasers (**a**) and Fabry-Perot type lasers (**b**)

arrange for some wavelength selectivity in the cavity configuration in order to maintain single-longitudinal-mode operation under dynamic conditions (modulation or temperature change). This laser is called a *Dynamic Single-Mode* (DSM) *laser* [9.4].

The following list includes several schemes for realizing a dynamic single-mode laser:

(i)   Distributed FeedBack (DFB).
(ii) Distributed Bragg Reflector (DBR).
(iii) Coupled cavity.
(iv) Short cavity.
(v)   Injection locking.

Let us explain some of these devices. First we consider the DFB laser. Since the resonant frequency is determined by the period of the grating prepared on the active layer or close to the active layer [9.5], it is likely to have a stable single-longitudinal-mode operation. Various schemes of DFB lasers have been reported. The GaAlAs/GaAs system was the first to be used for a DFB laser, and lasers of 1.3 μm and 1.55 μm have been put into practical use for optical communications. The DFB laser contains two equivalent resonant wavelengths for the lowest mode which require fabrication of an asymmetric structure along the axial direction or phase-adjustment at the central region, illustrated in Fig.9.6.

In the DBR laser, the grating is prepared on both sides or on one side of the active layer and acts as a reflecting mirror with frequency selectivity. When the grating is prepared on the same layer as the active region, there exists an appreciable amount of absorption so that high reflectivity cannot be achieved. For this reason some methods have been proposed for preparing the grating away from the active layer. An Integrated-Twin-Guide (ITG)

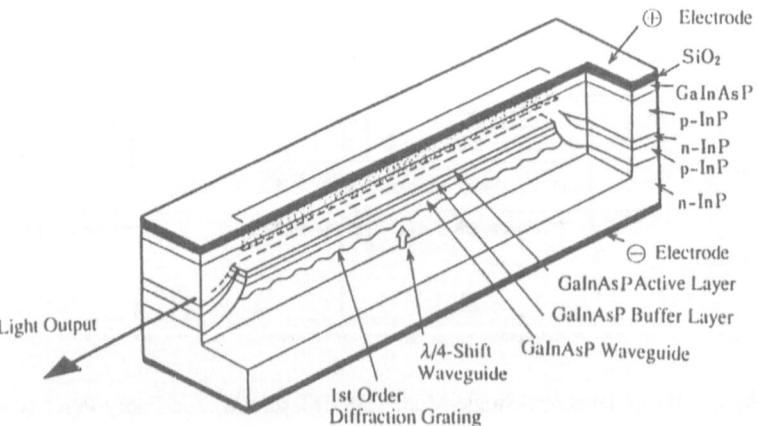

Fig.9.6 A phase-shift DFB laser [9.6]

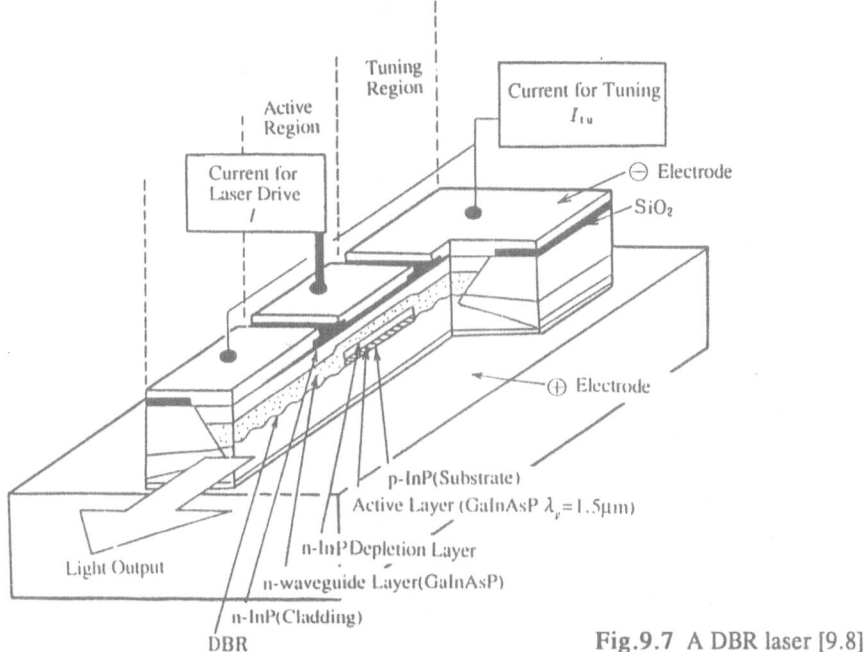

Fig.9.7 A DBR laser [9.8]

structured laser [9.7] has a single-mode operation without mode hopping even when modulated at 1 GHz or higher. Figure 9.7 schematically depicts that the DBR laser has an output waveguide that is directly coupled to an active waveguide, has a high coupling efficiency, and is appropriate for optical integrated circuits due to the low loss of the waveguide. These important qualities have generated intensive research efforts directed toward DBR lasers.

In DFB and DBR lasers, the wavelength variation against a change of temperature is as small as approximately 1 Å/K. We stress again the fact that a single longitudinal mode can be maintained even when the laser is modulated by a high-frequency signal. It is very important for use as a light source in the 1.5 to 1.6 μm wavelength region where ultra-low-loss transmission is possible, but in any case a laser with narrow linewidth is necessary for wide-band communication. In the InGaAsP/InP system, it has been revealed that the device lifetime is not as short as that of the GaAlAs/GaAs system.

Table 9.1 compares various lasers which can become DSM lasers. DFB lasers of 1.3 μm have been used for high-speed transmission since 1988. However, the current type of laser does not meet the conditions stated above. The combination of an asymmetric phase-shift structure and a window structure, or DBR-type lasers still have future possibilities and the combination with an external modulater renders them very effective.

Table 9.1. Classification of dynamic single-mode (DSM) lasers

Single Mode

Structure \\ Requirement	DFB DBR	Short Cavity		Compund Resonator	Injection Locking
		Etching	Surface Emitting		
Mass Production	O	O	O	O	×
Wafer Test	O△	O	O	△	×
Integration	O	O	O	△	×
2D Array	×	△	O	×	×
Vertical Emission	×△	△	O	×	×

Recently, in order to improve the lasing characteristics of conventional DFB and DBR lasers, a Distributed-Reflector (DR) laser has been proposed and demonstrated [9.9]. In DFB and DBR lasers, it is difficult to achieve high-power operation, high device efficiency and good dynamic single-mode properties. The DR-laser structure, consisting of active and passive distributed reflectors, as illustrated in Fig.9.8, is able to operate at high

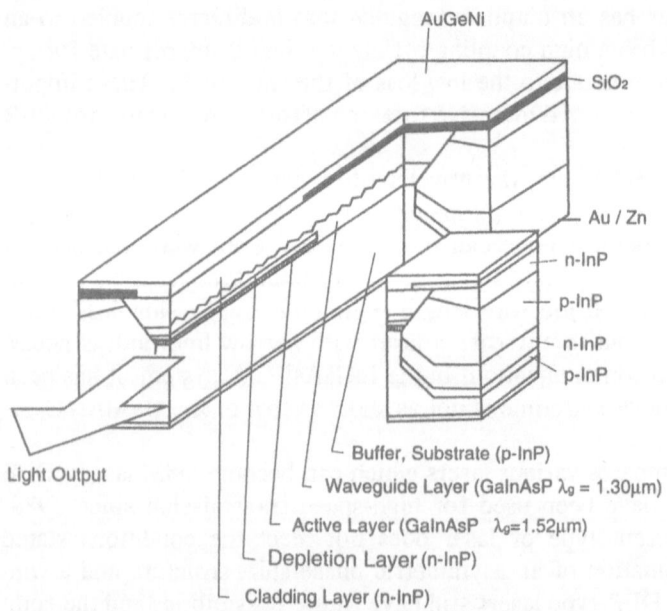

Light Output

AuGeNi
SiO₂
Au / Zn
n-InP
p-InP
n-InP
p-InP
Buffer, Substrate (p-InP)
Waveguide Layer (GaInAsP λ$_g$ = 1.30µm)
Active Layer (GaInAsP λ$_g$=1.52µm)
Depletion Layer (n-InP)
Cladding Layer (n-InP)

Fig.9.8. A Distributed Reflector (DR) laser [9.9]

output power and high output efficiency without degrading the spectral properties of conventional DSM lasers. Moreover, the DR laser has an advantage over DFB lasers because of its capability for polarization stabilization by loading a TM-mode filter. The 1.5 μm GaInAsP/InP BIG-DR laser obtained single-facet light output power with more than 90 % total power with high yield under single-mode operation.

9.3 Burying Epitaxy on Mesas and V-Grooves

9.3.1 Structures of Index-Guided Lasers

In order to realize a transverse-mode-controlled structure, epitaxial growth on mesas or V-grooves is very important, yielding a buried-type or a selective-growth-type laser (Sect.9.1.1). Representative structures that have been realized are shown in Fig.9.9.

Figure 9.9a depicts a structure in which an optical waveguide function is provided in the direction horizontal to the active layer, causing a transverse mode to remain stable even when the current is increased, and a longi-

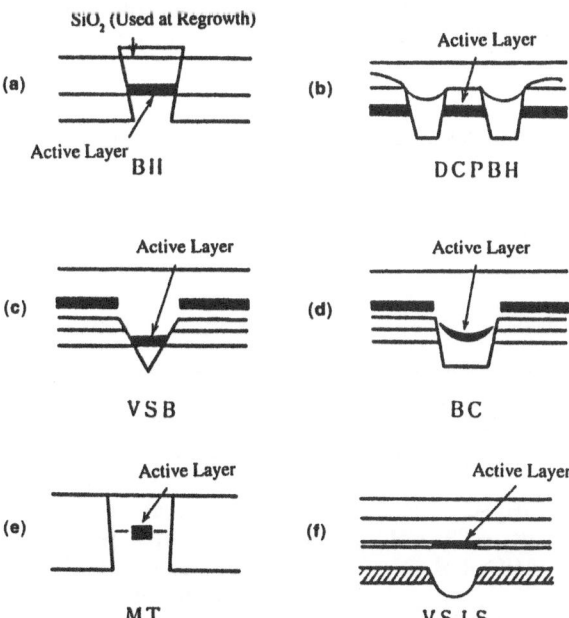

Fig.9.9a-f. Representatives of index-guided lasers (BC: buried crescent type, BH: buried heterostructure, DCPBH: double channel planar buried heterostructure, MT: mass-transport buried type, VSB: V-grooved substrate buried type, VSIS: V-grooved substrate internal stripe type)

tudinal mode is made single in most cases. In Fig.9.9b, an optical wave-guide function in the horizontal direction (layer direction) is realized by an equivalent refractive-index waveguide. Both the optical-confinement effect due to the difference of the refractive indices in the horizontal direction and a gain-guide function at the time of the current increase are provided. In Fig.9.9c, the structure has a waveguide function corresponding to the difference in impurity concentration.

Epitaxial growth around mesas or on grooves formed by LPE is faster than the flat surface and the growth does not occur on dielectric films. These features easily form a waveguide structure and flatten the surface. The fabrication procedure for index-guided lasers which make the most of these features is explained below.

9.3.2 Fabrication of Transverse-Mode-Controlled Structures

Ridges and mesas which are used in Burried Heterostructures (BH) lasers and the associated etchants are summarized in Table 9.2. An easily fabri-cated (one-step-grown) BH structure can be achieved if a double hetero-structure is grown on the mesa formed in advance on the substracte (1) and (3) in the figure.

Concerning the types (2) and (4), the double heterostructure is grown on the substrate before forming the mesa. Mesas are formed by etching the wafer to the bottom of the active layer or more, and burying layers are grown around the mesa during the second step of epitaxial growth.

Groove and terraced structures and the proper etchants are listed in Table 9.3. Crystal growth on grooves or near the side wall of a terraced structure occurs promptly, so that thicker layers can automatically be grown on grooves or near the side wall.

Table 9.2. Ridge and mesa etchants

Material	Structure	Etchant	Reference
GaAs	(1) ⊓	$8H_2SO_4 + H_2O_2 + H_2O$	*Iida* et al. [9.10]
GaAlAs/GaAs	(2) ⊓	$Br_2 + CH_3OH$	*Tarui* et al. [9.11]
InP	(3) ⊓	$4HCl + H_2O$	*Kambayashi* et al. [9.12]
			Kishino et al. [9.13]
InP/GaInAsP	(4) ⊓	$HCl, 3H_2SO_4 + H_2O_2 +$	
		H_2O	*Hsieh* et al. [9.14]
		$Br_2 + CH_3OH$	*Itaya* et al. [9.15]

Table 9.3. Groove and terraced structures and etchants

Material	Structure	Etchant	Reference
GaAs		$50\,H_2O + H_2O_2 + NH_4OH$	*Yamamoto* et al. [9.16]
		20 ethylene glycol +	
		$5H_3PO_4 + H_2O_2$	*Burmham* et al. [9.17]
		----------	*Sugino* et al. [9.18]
InP		$3HCl + H_3PO_4$	*Ishikawa* et al. [9.19]
		$5H_2SO_4 + H_2O_2 + H_2O$	*Doi* et al. [9.20]
		$HCl + 2CH_3COOH +$	
		H_2O_2 (KKI)	*Moriki* et al. [9.21]

Selective growth is possible by the use of a dielectric film. Placing a dielectric film on the mesa top prevents crystal growth onto the mesa. Then a current-blocking layer can be formed with the exception of the mesa, which is assigned to a current channel.

On the other hand, the same structure can be made without any dielectric mask at a particular growth condition. Figure 9.10 depicts the cross section of a DCPBH laser [9.22]. In this structure, the current-blocking structure is formed by one of the features of LPE. When the width of the mesa decreases, the crystal growth on the mesa top becomes thinner. Under a certain condition, no crystal growth occurs at last without a SiO_2 mask on the mesa top. Then the current-blocking layer can be formed with the exception of the mesa at the second crystal growth.

As in the case of a S-MIS structure, mentioned in Sect.9.1.2, and a of CSP structure [9.3], the equivalent index difference along the horizontal direction occurs due to the thickness difference of the layer just on the V-groove. This is in spite of a uniform thickness of the active layer, which works as a transverse-mode-controlled structure.

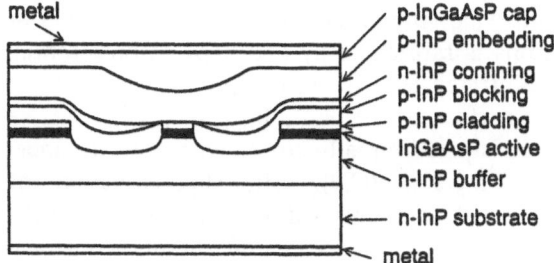

Fig.9.10. A Double-Channel Planar Buried Heterostructure (DCPBH) laser [9.22]

(a) Double heterostructure laser wafer

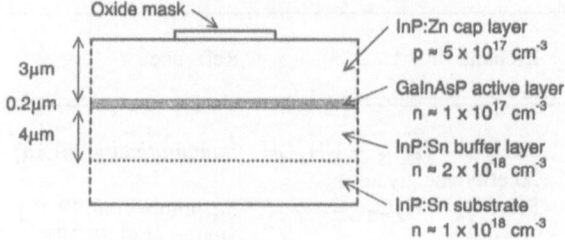

3μm

0.2μm

4μm

Oxide mask

InP:Zn cap layer
p ≈ 5×10^{17} cm^{-3}

GaInAsP active layer
n ≈ 1×10^{17} cm^{-3}

InP:Sn buffer layer
n ≈ 2×10^{18} cm^{-3}

InP:Sn substrate
n ≈ 1×10^{18} cm^{-3}

(b) Selective chemical etching

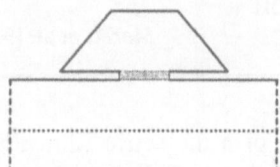

(c) Transport of InP

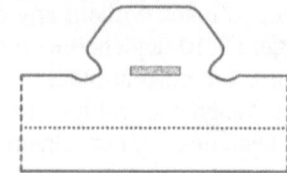

Fig.9.11. Mass transport procedure [9.23]

9.4 Mass-Transport Technique

The mass transport technique developed by *Liau* and *Walpole* [9.23] greatly simplifies the fabrication of GaInAsP/InP BH lasers. The phenomena is based on the mass transport through the vapor phase in an H_2 and PH_3 atmosphere. For example, this technique is applicable to the fabrication of BH lasers as follows: by the experiment depicted in Fig.9.11, the mesa stripe structure with $5 \div 6$ μm wide stripes parallel to either the (011) or (011) crystallographic directions is fabricated by two steps of preferential chemical etching. First, concentrated HCl is used to remove the InP cap layer. The etching stops at the surface of the $Ga_{0.27}In_{0.73}As_{0.63}P_{0.37}$ active layer. Next, the GaInAsP layer is preferentially etched with a 50 mℓ aqueous solution of 10 g KOH and 0.2 g $K_3Fe(CN)_6$. After slightly etching the mesa surface in buffered HF for 1 min, the wafer is placed into a LPE system. The wafer is loaded into the substrate holder of a graphite boat and is covered by a graphite cover, as mentioned in Sect.5.2.2. The wafer is heated while H_2 and PH_3 are flowing. Their flow rates are chosen so as not to affect the flat surface except for the mesa edges. The experimental system reaches 670°C in approximately 30 min and remains stable for 30 min before being rapidly cooled down.

The mesa shape is remarkably different after the mass transport process has taken place, as revealed. The mesa corners are eroded, while the narrow undercut channels are filled in with InP. The transport of InP has been reproducibly observed with the exception of when PH_3 was not flowing.

The mass transport phenomena can be thought to be due to the driving force necessary to minimize the surface energy. However, a detailed study of the phenomena has not yet been attempted. The mass transport technique is currently popular and has been widely used for DFB lasers [9.24, 25], a surface emitting laser with a 45° deflector [9.26], and for an InP lens formation [9.27]. The mass-transport technique for the GaAlAs/GaAs system is currently undeveloped.

9.5 Selective Meltback Technique

9.5.1 Selective Meltback Characteristics

The enhancement caused by dissolving an unsaturated melt on various materials (meltback characteristics) can also be effectively applied to structured epitaxial layers and is called the *selective meltback technique*. This technique was demonstrated by fabricating a GaAlAs/GaAs laser such as a selective meltbacked substrate innerstripe (S-MIS) laser [9.2] and a BH laser [9.28, 29]. Recently, the selective meltback technique has been instrumented for realizing low threshold surface emitting lasers [9.29, 30].

The Al-content dependence of the meltback-speed on the ⟨100⟩ direction, one of the selective meltback characteristics, is evaluated in detail by measuring the meltback depth of the $Ga_{1-x}Al_xAs$ layers with various Al contents x versus the meltback time. A series of $3 \div 10$ µm thick GaAlAs epitaxial layers with various Al contents and thin GaAs cap layers were grown on GaAs substrates as samples. The GaAs cap layer is used to prevent the GaAlAs layer from becoming oxidized. After fabricating a SiO_2 film on the wafer, a 450-µm pitch and 100-µm wide stripe window is formed by a photolithographic and chemical etching technique. These samples are melted back in a LPE furnace and the meltback characteristics are evaluated by measuring the depth of a dissolved groove. Figure 9.12 presents the meltback characteristic of the following meltback conditions:

T_M Meltback starting temperature (786°C)
R   Cooling rate (0.5°C/min.)
ΔT The degree of undersaturation for Ga and As in the solution, which is the temperature drop from that for saturation (8°C).

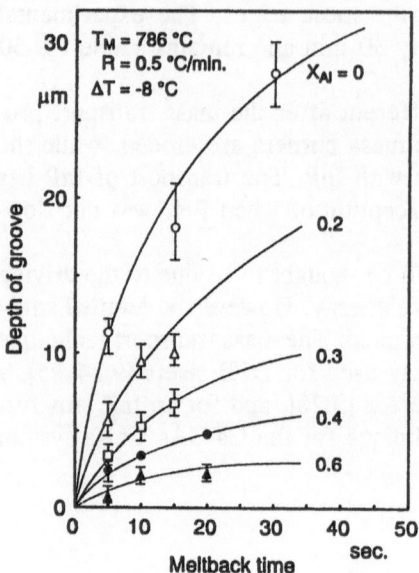

Fig.9.12. Selective meltback chracteristic with various Al contents

The measured results show a sublinear characteristic, and it can clearly be seen that the amount being melted back decreases with the increase of the Al content of the epitaxial layer. For instance, at 10 s of meltback time, the depths of the grooves are 16 μm for GaAs, 9 μm for $Ga_{0.8}Al_{0.2}As$, 5 μm for $Ga_{0.7}Al_{0.3}As$, 3 μm for $Ga_{0.6}Al_{0.4}As$, and 2 μm for $Ga_{0.4}Al_{0.6}As$, respectively. Under this condition, the selectivity is good over $5 \div 10$ s. The meltback speed can be regulated by changing ΔT.

9.5.2 Application to an Inner-Stripe Structure

The selective meltback technique is useful for fabricating V-groove structures. In this process, the formation of a V-grooved substrate and the following epitaxial growth are no longer separate processes but become the same in-situ process in the LPE furnace. The selective meltback technique therefore simplifies the fabrication and keeps the wafer from being polluted in the air during the process. Let us explain the minute details of the fabrication process applied to an inner-stripe laser.

The first step is the formation of a mesa stripe on the substrate, as depicted in Fig.9.13. The next process occurs in the LPE furnace. LPE epitaxial growth flattens the wafer surface so that the first epitaxial layer, a p-type $Ga_{0.5}Al_{0.5}As$ blocking layer, becomes thinner on the mesa top than on the remainder of the wafer, also illustrated in Fig.9.14. The wafer is melted in an unsaturated melt after the growth of the blocking layer. As the wafer is being melted, the GaAs mesa top emerges. Then the mesa is selectively

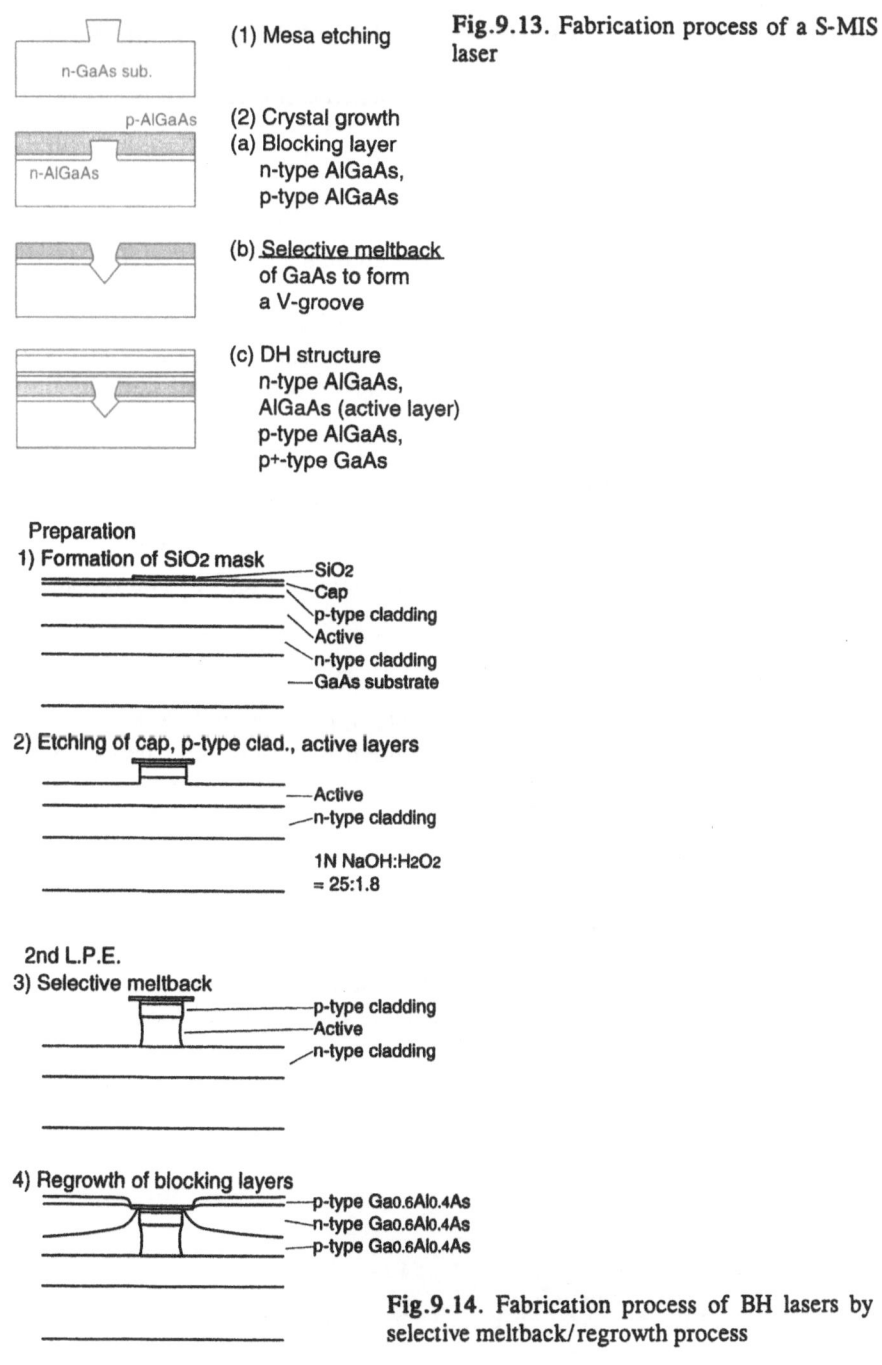

(1) Mesa etching

Fig.9.13. Fabrication process of a S-MIS laser

(2) Crystal growth
(a) Blocking layer
   n-type AlGaAs,
   p-type AlGaAs

(b) Selective meltback
   of GaAs to form
   a V-groove

(c) DH structure
   n-type AlGaAs,
   AlGaAs (active layer)
   p-type AlGaAs,
   p+-type GaAs

Preparation
1) Formation of SiO2 mask

2) Etching of cap, p-type clad., active layers

   1N NaOH:H2O2
   = 25:1.8

2nd L.P.E.
3) Selective meltback

4) Regrowth of blocking layers

Fig.9.14. Fabrication process of BH lasers by selective meltback/regrowth process

melted back because the meltback speed of GaAs is much faster than that of $Ga_{0.5}Al_{0.5}As$. The meltback on the mesa stops when the $(1, 1, 1)$ series crystal surfaces emerge. The meltback speed depends not only on material differences, but possibly on the crystal direction as well. This may be the cause of the formation of the V-groove. The double heterostructure is then successively grown on the V-grooved wafer. This in-situ process is one of the most attractive features and is different from the conventional method in which a V-groove is first formed by wet etching and then a double heterostructure is grown by the LPE process.

The threshold current of the primary experiment was between 65 to 73 mA at room temperature, CW conditions. A threshold current of 40 to 60 mA was also obtained when a p-type GaAs substrate was employed.

9.5.3 Application to BH Stripe Lasers

Selective meltback is also applicable to the fabrication of BH lasers. The selective meltback/regrowth process is illustrated in Fig.9.14 and is detailed as follows:

(i) Fabrication of a SiO_2 stripe mask on the DH wafer.
(ii) Preferential etching of p-type cap and p-type cladding layers.
(iii) The GaAs active layer (lower Al content material) with the exception of the mesa is selectively melted back in a LPE furnace to make a transverse-mode-controlled structure.
(iv) The p-, n- and p-type $Ga_{0.6}Al_{0.4}As$ current blocking layers are successively grown on both sides of the mesa stripe.

A cross-sectional view of the BH stripe laser is depicted in Fig.9.15. It is seen that the 16-μm wide and 3-μm thick active region which has a spool-like shape melted back from both sides is located at the center of the mesa

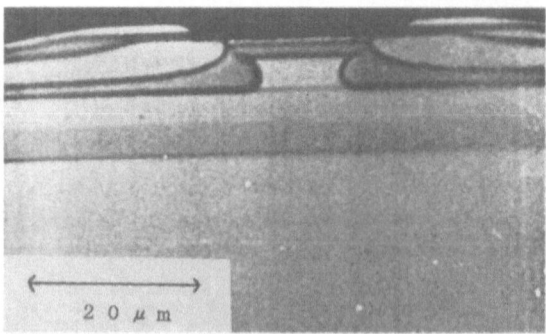

$2\,0\,\mu\,m$

Fig.9.15. Cross-sectional view of a BH stripe laser

stripe. The active layer is buried by the p-, n- and p-type GaAlAs blocking layers. The n-type cladding layer was not melted back and is almost flat. This means that the n-type cladding GaAlAs layer definitely performs as a stop layer to control the meltback depth. It is understood that the selective meltback/regrowth is an effective method for fabricating a transverse-mode-controlled structure, especially for a DH wafer which has a thick active layer.

9.6 Overgrowth on Gratings

Overgrowth on gratings is an important technique to realize a DSM laser. The height of the grating greatly affects the coupling efficiency. During the soak period in which source material and dopants are mixed in the melt, the corrugation deformation has been confirmed due to the high vapor pressure of phosphorus [9.31]. Recently, deformation due to the mass transport was also discussed. The following new methods have been proposed, capable of achieving hights of $400 \div 600$ Å: low temperature ($\approx 589°$C), flowing of PH_3 [9.32], and GaAs cover [9.33].

9.7 Growth of Quantum Wells

Let us assume that we are going to fabricate a potential well (Fig.9.16), by means of a heterojunction formed from two kinds of semiconductors with different band gaps. When electrons are put into the well, they occupy discrete energy levels in the well, as is well known from elementary quantum mechanics. In other words, in spite of a material being a semiconductor, the energy level which, at first, is a continuous energy band changes to possess discrete energy levels by being fabricated into a well-type potential. This phenomenon is called the *quantum-size effect*: the energy level changes by fabricating the artificial potential, and extensive research on this subject is being carried out in semiconductor engineering and physics. A semiconductor laser using a quantum well as a laser medium is called a *quantum-well laser* (Fig.9.16).

The wave function of an electron in the quantum well is expressed by the Schrödinger equation

$$\left[-\frac{\hbar^2}{2m_e^*} \frac{d^2}{dz^2} + U(z) \right] \psi(z) = E \psi(z) \tag{9.10}$$

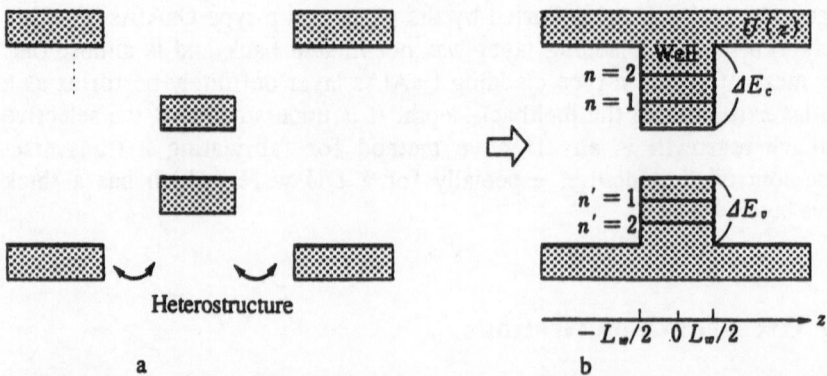

Fig.9.16a,b. Quantum-well structure: **(a)** before the formation of the junction and **(b)** thereafter

where $U(x)$ is a well-type potential, and E_n is an eigenvalue of energy to be obtained later, and $\hbar$ is Planck's constant divided by 2π, and m_e^* is the electron's effective mass given in (2.1). The solution of (9.10) becomes

$$\psi(z) = \begin{cases} A\begin{bmatrix} \cos \\ \sin \end{bmatrix}(\kappa z) & (|z| \leq \tfrac{1}{2}L_w) \\[2ex] A\begin{bmatrix} \cos \\ x/|x| \cdot \sin \end{bmatrix}(\kappa L_w/2)\exp[-\gamma(|z|-\tfrac{1}{2}L_w)] & (|z| > \tfrac{1}{2}L_w) \end{cases} \qquad (9.11)$$

where $\kappa = \hbar^{-1}(2m_e^* E)^{1/2}$ and $\gamma = \hbar^{-1}[2m_e^*(\Delta E_c - E)]^{1/2}$. The upper equations correspond to an even mode and the lower ones to an odd mode.

From the condition that a wave function and its first-order derivative are continuous at the boundary of a well, the following equation can be obtained as a characteristic function for the energy E_N

$$V = \frac{1}{\sqrt{1-b}}\left[\tan^{-1}\sqrt{\frac{b}{1-b}} + N\frac{\pi}{2}\right] \quad (N = 0, 1, 2...) \qquad (9.12)$$

$$V = \frac{L_w}{\hbar}\sqrt{2m_e^*\Delta E_c}\ , \quad b = 1 - E_N/\Delta E_c\ . \qquad (9.13)$$

Conventionally, the mode number n is employed as $n = N+1$ with $n = 1, 2, 3,$.

Figure 9.17 plots the results from a numerical solution of (9.12, 13) numerically. Thus far, only the vertical motion to the connecting surface of the well in Fig.9.16 has been discussed. However, electrons can also move in the direction parallel to the well, and so the energy is written as

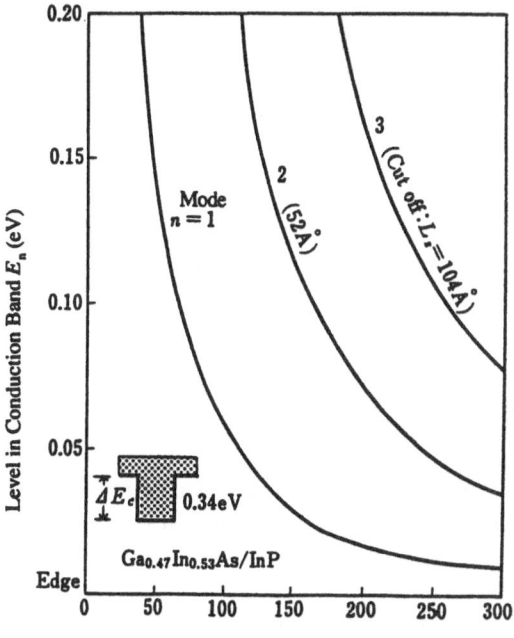

Well Width L_z(Å)

Fig.9.17. Energy levels in a quantum well

$$E = E_n + \frac{\hbar^2 k_{\parallel}^2}{2m^*} , \qquad (9.14)$$

$$m^* = \frac{m_e^* m_h^*}{m_e^* + m_h^*} \qquad (9.15)$$

where $\hbar k_{\parallel}$ is the momentum parallel to the junction of the well, and m^* is the reduced mass. Therefore, the second term in (9.14) expresses the kinetic energy.

Because energy becomes discrete, the density of states of the electrons is plotted in Fig.9.18 and written as

$$g_{cv} = \sum_{n=1}^{\infty} \frac{1}{2\pi L_w} \frac{2m^*}{\hbar^2} H(E - E_n) \qquad (9.16)$$

where $H(x)$ is Heaviside's function:

$$H(x < 0) = 0 \quad \text{and} \quad H(x \geq 0) = 1 . \qquad (9.17)$$

143

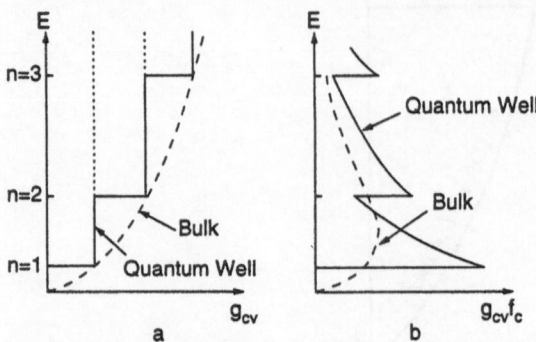

Fig.9.18a,b. Energy diagram of a quantum well and bulk crystal. (a) Density of states, $g_{cv}(E)$, (b) carrier distribution $g_{cv}(E)f_c(E)$

It should be noted that the density of states becomes stepwise and the spread of electrons becomes narrow. Consequently, as is depicted in Fig.9.19, the gain becomes larger and its profile sharper than those of ordinary bulk semiconductors. The quantum-well laser features an ultralow threshold, the ability to specify the oscillation wavelength by chosing the proper well

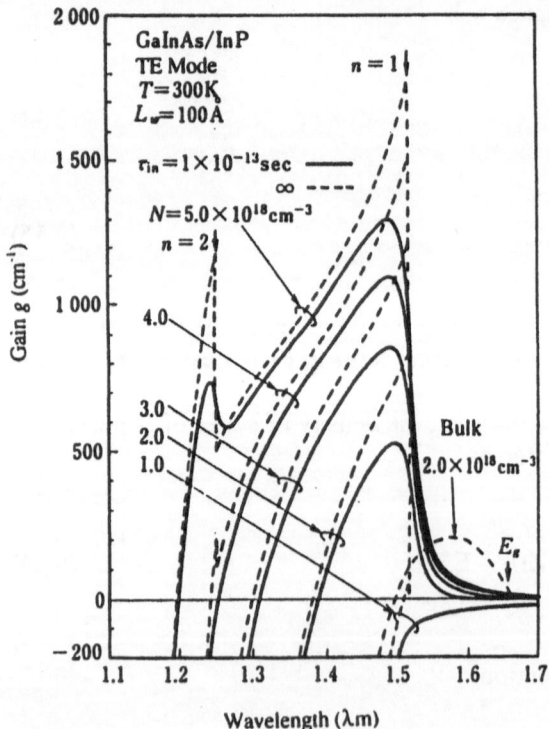

Fig.9.19. Gain spectra of a quantum-well laser [9.34]

thickness, excellent temperature characteristics, and a narrow spectral width of the gain. Intensive research and the development of quantum-well lasers continues to the present day; a device with a threshold of less than 1 mA has been realized.

The important factors of crystal growth for a quantum well are the thickness controllability of an epitaxial layer on the order of 10 Å, and the sharpness of the hetero-interface.

Sasai et al. [9.35] employed a low-temperature LPE technique at 589 °C to fabricate Multi Quantum-Well (MQW) lasers which offers a smaller growth rate and less meltback effect. They used a horizontal boat in which the substrate quickly moved back and forth between two barrier-layer melts arranged at both sides of the well layer melt. The growth time of the well layer was about $0.1 \div 0.2$ s. The epitaxial wafer consisted of a Te-doped n-type InP buffer layer (n $= 1 \cdot 10^{18} \text{cm}^{-3}$), five undoped 1.3 μm InGaAsP well layers ($L_z \simeq 200 \text{Å}$), undoped 1.1 μm InGaAsP barrier layers (d $\simeq 400 \div 600 \text{Å}$), a Zn-doped p-type InP cladding layer (p $= 5 \cdot 10^{17} \text{cm}^{-3}$), and a Zn-doped p-type InGaAsP contact layer (p $\simeq 2 \cdot 10^{18} \text{cm}^{-3}$).

The size of the fabricated MQW lasers included a $2 \div 3$ μm stripe width and a cavity length of about 250 μm. The threshold of the laser was 15 mA and the measured T_0 value was $130 \div 145$ K in the temperature range of less than 300 K, which originates from the quantum-size effect. These high-T_0 value were in good agreement with those obtained on 1.16-μm GaInAsP MQW lasers and the theoretical values predicted by *Dutta* et al. [9.36]. They also confirmed the size effect through a difference in gain between TE and TM modes due to the selection rule for optical transitions. They have obtained a larger gain difference of 65 cm^{-1} than that of the conventional DH lasers (10cm^{-1}) by the measurement of the gain-current relationship in spontaneous emission.

9.8 Growth of Multilayer Bragg Mirrors

The GaInAsP/InP hetero multilayer Bragg reflector was grown by the 2-phase solution LPE technique and by a computer-controlled rotational graphite boat system [9.37]. To facilitate use of the mirror for a 1.5 μm surface emitting laser (whose wavelength is 1.42 μm at 77K), the compositions of the $Ga_x In_{1-x} As_y P_{1-y}$ quaternary layers were selected to be x = 0.28 and y = 0.62. The associated band-gap wavelength and the refractive index were $\lambda_g = 1.3$ μm and $n_1 = 3.45$ (at $\lambda = 1.42 \mu$m), respectively. The thickness of each layer was maintained by controlling its growth time. The growth time for the GaInAsP layers was kept at 1.1 s each, but for the InP layers it varied from 0.8 s in the first layer to 2.2 s in the last layer because the growth rate of InP decreased when the layer number increased. Using this growth schedule, 50 layers of GaInAsP/InP hetero multilayers with one quarter-wavelength thickness were fabricated. The designed thickness of the InP layers (refractive index $n_2 = 3.19$ at $\lambda = 1.42\,\mu m$) was 1100 Å and for the GaInAsP layers was 1010 Å. The thickness of the grown InP layer was about $1100 \div 1200$ Å, and the GaInAsP layer was about 1000 Å. The measured reflectivity at the center wavelength of 1.4 μm was approximately 82%. Compared with the theoretically expected value of 96%, 82% was slightly smaller due to fluctuations in the thickness and absorption of each layer.

10. Surface-Emitting Lasers

The surface-emitting laser began to be considered one of the crucial devices for the next generation of optoelectronics when advantage was taken of its high degree of parallelism. In this chapter, we review the concept of a vertical-cavity surface-emitting laser, mainly consisting of the GaInAsP/InP, GaAlAs/GaAs, and GaInAs/GaAs systems. Their ultimate performance and impact on the electronics industry will be discussed.

10.1 The Concept of Surface-Emitting Lasers

The importance of ultrahigh-performance semiconductor lasers is rapidly increaasing along with the progress of large-capacity parallel light-wave communications, multi-access optical discs and optical computing. However, the present structure of cleaved semiconductor lasers still presents some problems. Among the difficulties yet to be solved, the following are included: the initial probe test of such devices is impossible before separating the wafer into chips, and the monolithic integration of lasers into optical circuits is limited due to the finite cavity length.

The Fabry-Perot resonator for a vertical-cavity Surface-Emitting (SE) laser is formed by two surfaces of an epitaxial layer with the light output emerging vertically from the surface [10.1]. This laser-structure scheme, if realized, would provide many advantages [10.2]; i.e., (i) the laser device is fabricated by a fully monolithic process, (ii) a densely packed two-dimensional laser array could be fabricated, (iii) the initial probe test could be performed before separation into chips, (iv) dynamic single-longitudinal-mode operation is expected because of its large mode spacing ($= 100 \div 200 \text{Å}$), (v) it is possible to vertically stack multiple-thin-film functional optical devices onto the SE laser, and (vi) a sharp, circular beam is achievable.

10.2 Structure and Characteristics

10.2.1 GaInAsP/InP Surface-Emitting Lasers

A GaInAsP/InP SE laser with a Circular Buried Heterostructure (CBH), strictly speaking, with a Flat-surface CBH (FCBH) has been fabricated [10.3]. This laser wass manufactored by means of a successive, fully monolithic fabrication process utilizing a three-step Liquid-Phase Epitaxy (LPE) growth. The substrate is polished to 150 μm thickness, and on the n-type side a Au/Sn electrode is formed. Next, the substrate and etch stop layer are selectively etched off to form a short cavity (7μm). The p-type side Au/Zn/Au electrode is then formed. Finally, a SiO_2/Au reflector is costructed by evaporation on the bottom of an etched well only.

The current/light output characteristic of a CBH SE laser device at 77 K under CW conditions has been measured [10.4]. The minimum threshold value was 6 mA. Pulsed operation has been obtained [10.5]; and CW operation has been achieved at room temperature for 1.3 μm devices.

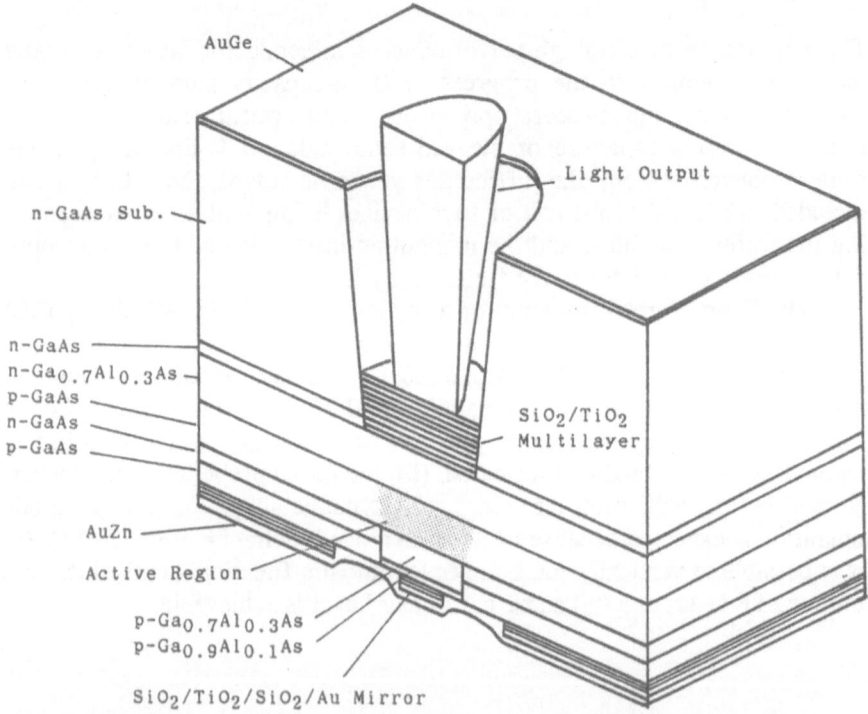

Fig.10.1. Schematic view of a Metal Organic Chemical Vapor Deposition (MOCVD) grown Circular Buried Heterostructure (CBH) GaAlAs/GaAs surface emitting laser

10.2.2 GaAlAs/GaAs SE Lasers

The GaAlAs/GaAs laser employs almost the same CBH structure as the GaInAsP/InP laser. In order to decreaase the threshold, the active region was constricted using the selective meltback method. A threshold of 6 mA was demonstrated for the active-region diameter of ≈ 6 μm under pulsed operation at 20.5°C [10.6]. It is notable that a micro-cavity of 7-μm length and 6 μm in diameter was achieved.

Figure 10.1 schematically depicts an MOCVD grown CBH SE laser [10.7, 8]. This laser was fabricated with a two-step MOCVD growth and utilizing entire-monolithic technology. First a GaAlAs/GaAs DH wafer with an active layer thickness of 3 μm was grown by MOCVD at 780°C under atmospheric pressure. After the first growth, a silicon nitride ($Si_3 N_4$) circular mask with a diameter of 10 μm was formed on the wafer for the mesa etch and the selective regrowth of current blocking layers. A p-type cladding GaAlAs layer was lightly etched using sulphuric acid ($H_2 SO_4$):$H_2 O$: hydrogen peroxide ($H_2 O_2$) = (1:8:8) solution. Selective MOCVD regrowth of GaAs under atmospheric pressure formed the current blocking layers (0.7μm thick n-type GaAs and 0.3μm thick p-type GaAs). The growth condition was the same as that used in the Double Heterostructure (DH) wafer growth. There was no deposition on top of a circular mesa which was covered with a $Si_3 N_4$ mask. A short cavity structure with a cavity length of 6 μm was formed by removing the GaAs substrate. A ring electrode with an outer/inner diameter of 40 μm/ 10 μm was adopted and the Au/SiN mirror became the n-type side mirror.

Figure 10.2 displays a typical current/light output characteristic and lasing spectrum under CW conditions at 20°C [10.9, 10]. The lowest CW threshold current was 30 mA, the differential quantum efficiency was typically 10 %, and the maximum CW output power was about 2 mW. The saturation of the output power is due to the temperature increaase of the device. Stable single-mode operation was observed with neither sub-transverse modes nor other longitudinal modes, as plotted in the inset of Fig. 10.2. The spectral width above threshold was less than 1 Å, which is limited by the resolution of the spectrometer (in this case a Anritsu MS9001A). Later, a high-resolution measurement of the spectral linewidth will be presented. The mode spacing of this device was 170 Å, which is 50 times larger than that of a conventional stripe laser. A side-mode suppression ratio of 35 dB was obtained at I/I_{th} = 1.25. This is comparable to that of well-designed DBR or DFB dynamic single-mode lasers.

The linewidth was measured using a *standard delayed self-homodyne method* with a 4 km long single-mode fiber [10.11]. In this method the light to be measured is first divided by a beam splitter, then the propagation of one beam is delayed with respect to the other one, before both are mixed again. Two optical isolators with a total isolation of 60 dB were used to

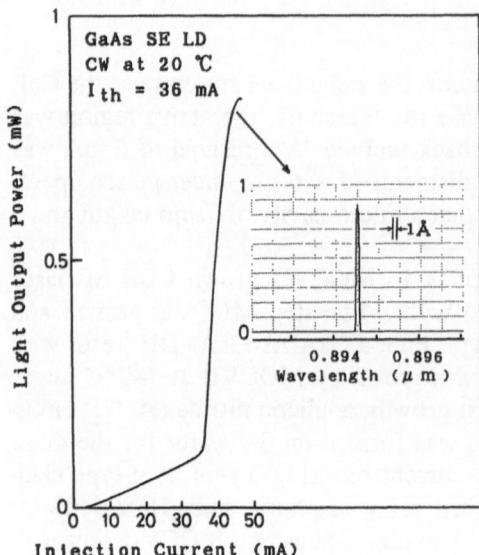

eliminate the effect of external optical feedback. The spectral linewidth of 50 MHz was obtained with an output power of 1.4 mW. Even in such an ultra-short cavity device with a length of less than 10 μm, a relatively narrow spectral linewidth was attained, attributable to the high reflectivity of the mirrors. A much narrower laser linewidth can be obtained by increasing the output power and reducing the cavity loss.

10.3 Semiconductor Multi-Layer Structure

Vertical-cavity SE lasers utilizing semiconductor multilayer reflectors for the DBR [10.12] or DFB [10.13] structures may enable the integration of thin-film functional optical devices onto a SE laser by stacking them. If this is achieved, new 3-dimensional integrated optics [10.14] will become available.

MOCVD, MBE, and CBE can easily provide the aforementioned superlattice structures, which enable DFB- and DBR-type SE lasers. For the purpose of realizing DBR SE lasers, DH wafers with a 3-μm thick GaAs active layer sandwiched by a couple of periodically layered Bragg reflectors composed of 30 layers of alternating GaAlAs and AlAs with quarter-wavelength are grown by MOCVD and CBE. A cross-sectional Scanning Electron Microscope (SEM) photograph of a MOCVD-grown active layer sandwiched by the multilayers and a part of the multilayer Bragg reflector are depicted in Fig. 10.3. The period of the Bragg reflector is 1400 Å. The reflectivity of

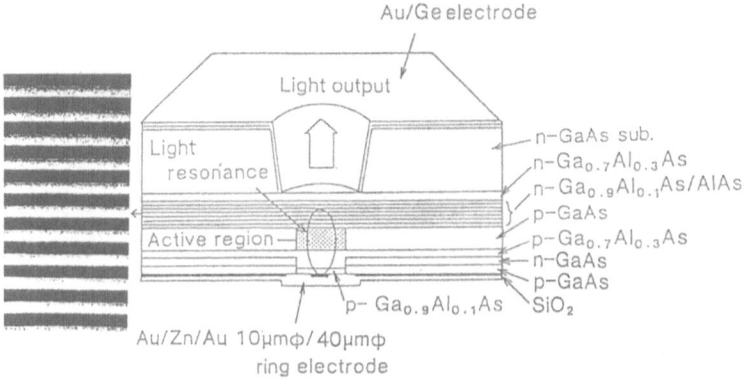

Fig.10.3. A cross-sectional Scanning Electron Microscope (SEM) photo of a wafer with a $Ga_{0.9}Al_{0.1}As/AlAs$ multi-layer Bragg reflector

the multilayer, Bragg reflector was measured from the top of the crystal surface. A maximum reflectivity of 97% was achieved at a wavelength of 0.87 μm, corresponding to the lasing wavelength of a GaAlAs/GaAs SE laser. Also, we have found that it is possible to inject carriers into an active region through multilayers by appropriately doping with impurities. We obtained for the first time oscillation of a GaAlAs surface-emitting laser which uses a multilayer reflector for one of the mirrors [10.12]. By introducing such a periodic configuration, a reduction in the threshold current can be expected [10.12].

A DBR or DFB structure without facet mirrors enables the integration of functional optical devices with SE lasers by stacking them. Once this structure has been realized, a new 3-dimensional integrated optics may become available.

10.4 Two-Dimensional Arrays

A conventional injection laser consists of two end mirrors cleved perpendicular to the active layer, so that only one-dimensional laser arrays can be monolithically fabricated; otherwise, the wafers must be stacked to form a two-dimensional laser array. However, it is possible to prepare a two-dimensional laser array by using SE-laser concepts. Specifically, a vertical-cavity SE laser can form high-density, two-dimensional arrays. Among the applications of such two-dimensional arrays is a high-power laser and stacked planar optics [10.14]. The concept of stacked planar optics is to construct a two-dimensional-array lightwave component by stacking two-dimensional

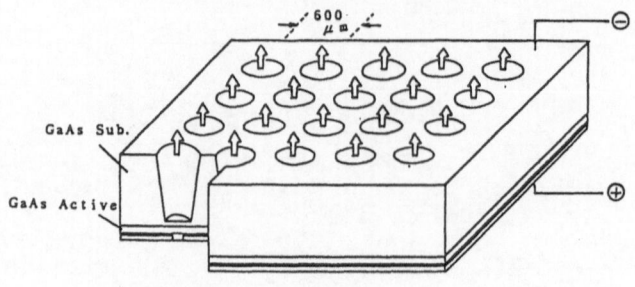

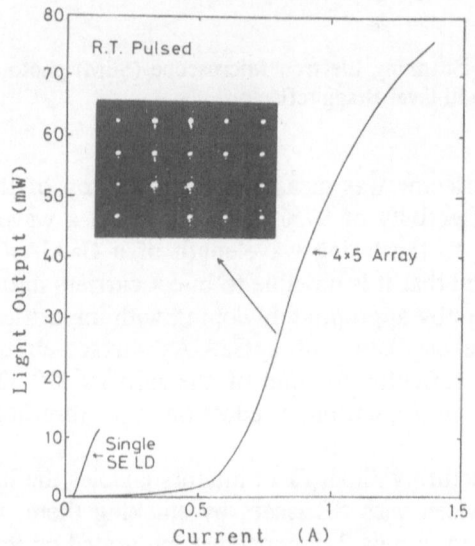

4 × 5 GaAs Two Dimensional Laser Array

Fig. 10.4. Two-dimensional surface-emitting laser array

planar optical device arrays with a planar microlens array. This configuration may enable the mass production of optical devices with easy alignment.

The first demonstration of a two-dimensional SE laser array included the GaInAsP system [10.15]. Another preliminary demonstration, a 4×5 GaAlAs/GaAs SE-laser array, was fabricated by a two-step MOCVD growth, detailed in Fig. 10.4 [10.2]. The separation of each device was 600 μm, and the current confining structure was the same as mentioned previously. This device operated under room-temperature pulsed conditions, providing 80 mW. Such high-density 2-D laser arrays can only be formed by utilizing a vertical-cavity configuration. We are also trying to coherently couple these arrayed lasers [10.6].

10.5 Ultralow-Threshold Devices

In order to reduce the threshold current of SE lasers, it is essential to increaase the mirror reflectivity and to reduce the device size. A lot of work has been done to make low-threshold SE lasers, such as using high-reflectivity semiconductor DBR [10.17-19], and introducing InGaAs strained quantum wells together with DBR [10.20,21]. With the progress of our SE laser research up to now, it is clear that the present performance of vertical-cavity SE lasers is not limited by any essential problem aside from technical hindrances. By overcoming technical problems such as the ohmic resistivity in the electrode and improving the heat sinking, we believe that we can obtain the 1 μA device depicted in Fig. 10.5.

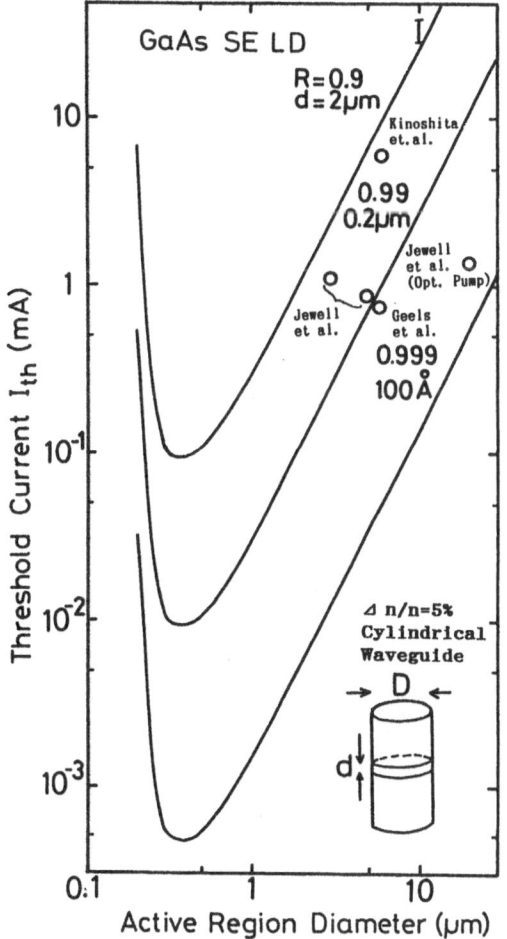

Fig. 10.5. Expected threshold currents of surface-emitting lasers

10.6 Future Prospects

If two-dimensional (2-D) array devices are available we can expect to simultaneously align a tremendous number of optical components, as in parallel multiplexing lightwave systems, parallel optical logic systems, high-power coherent laser arrays, etc. Stacking will permit a wide variety of functions, such as frequency tuning, amplification, and filtering to be integrated along with surface-emitting lasers.

A 2-D parallel optical logic system [10.20] can deal with a large amount of image information at high speed. To meet this demand, lasers will be key devices. Their technical aspects and future outlook may be found in [10.22]. Vertical optical interconnection of LSI chips and circuit boards may be another interesting topic of applied research. In any case, the 2-D dimensional array configuration of surface emitting lasers will lead into a new era of optoelectronics [10.23].

References

Chapter 1

1.1   H.K.V. Lotsch: Optik **26**, 112 and 181 (1967)
1.2   Y. Suematsu, K. Iga: *Introduction to Optical Fiber Communications*, 3rd edn. (Ohm-sha, Tokyo 1989) [Engl. Trans. of 1st edn. was published by Wiley, New York 1982]
1.3   H.C. Casey Jr., M.B. Panish: *Heterostructure Lasers* (Academic, New York 1978)
1.4   R.N. Hall, G.E. Fenner, J.D. Kingsley, T.J. Soltys, R.O. Carlson: Phys. Rev. Lett. **9**, 366 (1962)
1.5   T.M. Quist, R.H. Rediker, R.J. Keyes, W.E. Krag, B. Lax, A.L. McWhorter, H.J. Zeiger: Appl. Phys. Lett. **1**, 92 (1962)
1.6   M.I. Nathan, W.P. Dumke, G. Burns, F.H. Dill Jr., G. Lasher: Appl. Phys. Ltt. **1**, 62 (1962)
1.7   N. Holonyak Jr., S.F. Bevacqua: Appl. Phys. Lett. **1**, 82 (1962)
1.8   L. Solymar, D. Walsh: *Lectures on the Electrical Proprties of Materials*, 3rd edn. (Oxford Univ. Press, Oxford 1984) Sect. 9.19
1.9   T. Ikegami, Y. Suematsu: IEEE Proc., **55**, 122 (1967)
1.10 I. Hayashi, M.B. Panish, P.W. Foy, S. Sumski: Appl. Phys. Lett. **17**, 109 (1970)
1.11 Zh. I. Alferov, V.M. Andreev, D.Z. Garbuzov, Yu.V. Zhilyaev, E.P. Morozov, E.L. Portnoi, V.G. Trofim: Fiz. Tekh. Poluprov. **4**, 1826 (1970)
1.12 I. Hayashi, IEEE Trans. ED-31, 1630 (1984)
1.13 G. Lasher, F. Stern: Phys. Rev. A **133**, 553 (1964)
1.14 M.H. Pilkuhn, H. Rupprecht: Solid-State Electron. **7**, 905 (1964)
1.15 Y. Nannichi: J. Appl. Phys. **36**, 1499 (1965)
1.16 W.W. Anderson, IEEE J. QE-1, 228 (1965)
1.17 B.S. Goldstein, et al.: IEEE Proc. **53**, 195 (1965)
1.18 J. Takamiya, F. Kitasawa, J. Nishizawa: IEEE Proc. (Lett.) **56** 135 (1968)
1.19 T. Ikegami, Y. Suematsu: Trans. IECE Jpn. B **51**, 57 (1968); ibid. IEEE J. QE-4, 148 (1968)
1.20 J.C. Dyment: Appl. Phys. Lett. **10**, 84 (1967)
1.21 W. Susaki: Jpn. J. Appl. Phys. **6**, 977 (1967)
1.22 H. Kressel, H.P. Mierop: J. Appl. Phys. **38**, 5419 (1967)
1.23 H. Haug: Phys. Rev. **184**, 338 (1969)
1.24 N.G. Basov, V.V. Nikitin, A.S. Semenov: Usp. Fiz. Nauk **97**, 561 (1969) [Sov. Phys.-Uspekhi **12** 219 (1969)]
1.25 H. Kressel, H. Nelson: RCA Rev. **30**, 106 (1969)
1.26 J.E. Ripper, T.L. Paoli: Appl. Phys. Lett. **18**, 466 (1971)
1.27 T. Ikegami, K. Kobayashi, Y. Suematsu: Trans. IECE Jpn. B **53**, 53 (1970)
        T. Ikegami, Y. Suematsu: Trans. IECE Jpn. B **53** 513 (1970)
1.28 T.P. Lee, R.H.R. Roldan: IEEE J. QE-6, 339 (1970)
1.29 C.J. Hwang: Phys. Rev. B **2**, 4117 and 4126 (1970)

1.30 H. Kroemer: IEEE Proc. **51**, 1782 (1963)

1.31 W. Susaki, T. Sogo, T. Oku: IEEE J. QE-**4**, 422 (1968)

1.32 I. Hayashi, M.B. Panish, F.K. Reinhart: J. Appl. Phys. **42**, 1929 (1971)

1.33 T. Ikegami: IEEE J. QE-**8**, 470 (1972)

1.34 M.B. Panish, H.C. Casey Jr., S. Sumski, P.W. Foy: Appl. Phys. Lett. **22**, 590 (1973)

1.35 Y. Suematsu, M. Yamada: Trans. IECE Jpn. C **57**, 434 (1975)

1.36 K. Peterman: 7th Europ. Conf. Optical Commun. No.10.1 (1981)

1.37 W.T. Tsang, R.A. Logan, J.A. Ditzenberger: Electron. Lett. **18**, 845 (1982)

1.38 D.R. Scifres, C. Lindstrom, R.D. Burnham, W. Streofer, T.L. Paoli: Electron. Lett. **19**, 169 (1983)

1.39 Y. Suematsu: IEEE Proc. **71**, 692 (1984)

1.40 I. Hayashi: IEEE Trans. ED-**31**, 1630 (1984)

1.41 W.W. Chow, S.W. Koch, M. Sargent III: *Semiconductor-Laser Physics* (Springer, Berlin, Heidelberg 1994)

1.42 N.W. Carlson: *Monolithic Diode-Laser Arrays*, Springer Ser. Electron. Photon., Vol. 33 (Springer, Berlin, Heidelberg 1994)

Long-Wavelength Lasers

1.43 F.P. Kapron, D.B. Keck, R.D. Maurer: Appl. Phys. Lett. **17**, 423 (1970)

1.44 J.J. Hsieh, J.A. Rossi, J.P. Donnelly: Appl. Phys. Lett. **28**, 709 (1976)

1.45 A.P. Bogatov, L.M. Dolginov, P.G. Eliseev, M.G. Milvidskii, B.N. Sverdlov, E.G. Shevchenko: Sov. Phys. – Semicond. **9**, 1282 (1975)

1.46 K. Oe, K. Sugiyama: Jpn. J. Appl. Phys. **15**, 740 (1976)

1.47 T.P. Pearsall, B.I. Miller, R.J. Capik, K.J. Bachmann: Appl. Phys. Lett. **28**, 499 (1976)

1.48 T. Yamamoto, K. Sakai, S. Akiba, Y. Suematsu: Electron. Lett. **13**, 142 (1977)

1.49 K. Oe, S. Ando, K. Sugiyama: Jpn. J. Appl. Phys. **16**, 1273 (1977)

1.50 Y. Itaya, Y. Suematsu, K. Iga: Jpn. J. Appl. Phys. **16**, 1057 (1977)

1.51 K. Wakao, K. Moriki, T. Kambayashi, K. Iga: Jpn. J. Appl. Phys. **16**, 2073 (1977)

1.52 Zh. I. Alferov, A.T. Gorelenok, P. Kopiev, V.N. Mdivani, V.K. Tibilov: Pisma Zh. Tekh. Fiz. 3 1169 (1977)

1.53 M.A. Pollack, R.E. Nahory, J.C. DeWinter, A.A. Ballmann: Appl. Phys. Lett. **33**, 314 (1978)

1.54 T. Yamamoto, K. Sakai, S. Akiba, Y. Suematsu: IEEE J. QE-**14**, 95 (1978)

1.55 Y. Itaya, Y. Suematsu, S. Katayama, K. Kishino, S. Arai: Jpn. J. Appl. Phys. **18**, 1795 (1979)

1.56 S.D. Hersee, A.C. Carter, R.C. Goodfellow, G. Hawkins, I. Griffith: Solid-State Electron. Dev. **3**, 179 (1979)

1.57 R.J. Nelson: Appl. Phys. Lett. **35**, 654 (1979)

1.58 R.E. Nahory, M.A. Pollack: Electron. Lett. **14**, 727 (1978)

1.59 T. Yamamoto, K. Sakai, S. Akiba: IEEE J. QE-**15**, 684 (1979)

1.60 K. Mizuishi, M. Hirao, S. Tsuji, H. Sato, M. Nakamura: Jpn. J. Appl. Phys. **19**, L429 (1980)

1.61 H. Imai, M. Morimoto, H. Ishikawa, K. Hori, M. Takusagawa: Appl. Phys. Lett. **38**, 16 (1981)

1.62 E. Oomura, T. Murotani, H. Higuchi, H. Namizaki, W. Susaki: IEEE J. QE-**17**, 646 (1981)

1.63 A.G. Steventon, R.E. Spillet, R.E. Hobbs, M.G. Burt, P.J. Fiddyment, J.V. Collins: IEEE J. QE-17, 602 (1981)
1.64 K. Iga, B.I. Miller: Electron. Lett. 16, 830 (1980)
1.65 M. Hirao, S. Stujii, K. Mizushima, A. Doi, M. Nakamura: J. Opt. Commun. 1, 10 (1980)
1.66 J.J. Hsieh, C.C. Shen: Appl. Phys. Lett. 30, 429 (1977)
1.67 K. Kishino, Y. Suematsu, Y. Itaya: Electron. Lett. 15, 134 (1979)
1.68 K. Moriki, K. Wakao, M. Kitamura, K. Iga, Y. Suematsu: Jpn. J. Appl. Phys. 19, 2191 (1980)
1.69 H. Nishi, M. Yano, Y. Nishitani, Y. Akita, M. Takusagawa: Appl. Phys. Lett. 35, 232 (1979)
1.70 R.J. Nelson, P.D. Wright, P.A. Barnes, R.L. Brown, T. Cella, R.G. Sobers: Appl. Phys. Lett. 36, 358 (1980)
1.71 I. Mito, M. Kitamura, K. Kobayashi: Opt. Fiber Commun., Phoenix, AZ (1982) Paper ThBB2
1.72 T. Horikoshi, T. Kobayashi, Y. Furukawa: Jpn. J. Appl. Phys. 18, 2237 (1979)
1.73 D. Botez, G.J. Herskowitz: IEEE Proc. 68, 689 (1980)
1.74 M. Nakamura, S. Tsujii: IEEE J. QE-17, 994 (1981)
1.75 P. Marshall, E. Schlosser, C. Wolk: Electron. Lett. 15, 38 (1979)
1.76 S. Akiba, K. Sakai, T. Yamamoto: Jpn. J. Appl. Phys. 17, 1899 (1978)
1.77 S. Arai, Y. Suematsu, Y. Itaya: Jpn. J. Appl. Phys. 18, 709 (1979)
1.78 G.D. Henshall, P.D. Greene: Electron. Lett. 15, 621 (1979)
1.79 S. Akiba, K. Sakai, Y. Matsushima, T. Yamamolto: Electron. Lett. 15, 606 (1979)
1.80 H. Kawaguchi, T. Takahei, Y. Toyoshima, H. Nagai, G. Iwane: Electron Lett. 15, 669 (1979)
1.81 S. Arai, M. Asada, Y. Suematsu, Y. Itaya: Jpn. J. Appl. Phys. 18, 2333 (1979)
1.82 I.P. Kaminow, R.E. Nahory, M.A. Pollack. L.W. Stulz, J.C. DeWinter: Electron. Lett. 15, 763 (1979)
1.83 K. Iga, Y. Suematsu: 1st Europ. Conf. Integr. Opt., London (1981) p. 70
1.84 H. Kogelnik, C.V. Shank: Appl. Phys. Lett. 18 152 (1971)
1.85 I.P. Kaminow, H.P. Weber: Appl. Phys. Lett. 18, 497 (1971)
1.86 M. Nakamura, A. Yariv, H.W. Yen, S. Somekh, H.L. Garvin: Appl. Phys. Lett. 22, 515 (1973)
1.87 F.K. Reinhart, R.A. Logan: Appl. Phys. Lett. 27, 45 (1975)
1.88 M. Nakamura, K. Aiki, J. Umeda, A. Yariv: Appl. Phys. Lett. 27, 403 (1975)
1.89 H.C. Casey, Jr., S. Somekh, M. Ilegems: IEEE Trans. ED-22, 1060 (1975)
1.90 D.R. Scifres, R.D. Burnham, W. Streifer: Appl. Phys. Lett. 27, 295 (1975)
1.91 W. Tsang, S. Wang: 9th Int'l Quantum Electron. Conf. (Kanazawa 1976) p. 38
1.92 H. Kawanishi, Y. Suematsu, Y. Itaya, S. Arai: Jpn. J. Appl. Phys. 17, 1439 (1978)
1.93 A. Doi, T. Fukuzawa, M. Nakamura, R. Ito, K. Aiki: Appl. Phys. Lett. 35, 441 (1979)
1.94 Y. Suematsu, M. Yamada, K. Hayashi: IEEE Proc. 63, 208 (1975)
1.95 F.K. Reinhart, R.A. Logan, C.V. Shank: Appl. Phys. Lett. 27, 45 (1975)
1.96 C.E. Hurwitz, J.A. Ross, J.J. Hsieh, C.M. Wolf: Appl. Phys. Lett. 27, 241 (1975)
1.97 K. Aiki, M. Nakamura, J. Umeda: Appl. Phys. Lett. 29, 506 (1976)

Short-Wavelength Lasers

1.98 R.D. Burnham, N. Holonyak Jr., D.L. Keune, D.R. Scifres: Appl. Phys. Lett. 18, 160 (1971)

1.99 Zh. I. Alferov, I.N. Arsent'ev, D.Z. Garbuzov, S.G. Konnikov, V.D. Rumyantsev: Sov. Tech. Phys. Lett. 1, 147 (1975)

1.100 H. Kressel, G.H. Olsen, C.J. Nuese: Appl. Phys. Lett. 30, 249 (1977)

1.101 T. Mukai, H. Yajima, J. Shimada: Jpn. J. Appl. Phys. 20, L729 (1981)

1.102 T. Suzuki, I. Hino, A. Gomyo, K. Nishida: Jpn. J. Appl. Phys. 21, L731 (1982)

1.103 Y. Kawamura, H. Asahi, N. Nagai, T. Ikegami: Electron. Lett. 19, 931 (1983)

1.104 A. Fujimoto, H. Watanabe, M. Takeuchi, M. Shimura: Jpn. J. Appl. Phys. 23, L720 (1984)

1.105 A. Usui, T. Matsumoto, M. Inai, I. Mito, K. Kobayashi, H. Watanabe: Jpn. J. Appl. Phys. 24, L163 (1985)

1.106 K. Kishino, Y. Kaneka, A. Harada: Jpn. J. Appl. Phys. 24, L358 (1985)

1.107 K. Kobayashi, S. Kawata, A. Gomyo, I. Hino, T. Suzuki: Electron. Lett. 21, 931 (1985)

1.108 M. Ikeda, Y. Mori, H. Sato, K. Kaneko, N. Watanabe: Appl. Phys. Lett. 47, 1027 (1985)

1.109 M. Ishikawa, Y. Ohba, H. Sugawara, M. Yamamoto, T. Nakanishi: Appl. Phys. Lett. 48, 207 (1985)

1.110 J.P. Andre, E. Dupont-Nivet, D. Monori, J.N. Patillon, M. Erman, T. Ngo: J. Cryst. Growth 77, 354 (1986)

1.111 K. Nakano, M. Ikeda, A. Toda, C. Kojima: Electron. Lett. 23, 894 (1987)

1.112 K. Iga, H. Uenohara, F. Koyama: Electron. Lett. 22, 1008 (1986)

1.113 K. Kishino, A. Kikuchi, Y. Kaneka, I. Nomura: Appl. Phys. Lett. 58, 1822 (1991)

1.114 J. Rennie, M. Okajima, M. Watanabe, G. Hatakoshi: 13th IEEE Semicond. Laser Conf., Takamatsu (1992) Tech. Digest. No.G-4, 158

1.115 T. Kamizono, S. Arimoto, H. Watanabe, K. Kodoiwa, E. Omura, S. Kakimoto, K. Ikeda: 13th IEEE Semicond. Laser Conf., Takamatsu (1992) Tech. Digest. No.D-30, 94

1.116 H. Hamada, K. Tominaga, M. Shono, S. Honda, K. Yodoshi, T. Yamaguchi: Electron. Lett. 28, 1834 (1992)

1.117 A. Vlaster, C.J. van der Poel, M.N. Finke, M.J.B. Boermanns: 13th IEEE Semicond. Laser Conf., Takamatsu (1992) Tech. Digest No. G-1, 152

1.118 R.G. Hunsperger: *Integrated Optics: Theory and Technology*, 4th edn. (Springer, Berlin, Heidelberg 1995)

1.119 T. Tamir (ed.): *Guided-Wave Optolectronics*, 2nd edn., Springer Ser. Electron. Photonics, Vol.26 (Springer, Berlin, Heidelberg 1990) Chap.5

1.120 H. Kressel (ed.): *Semiconductor Devices for Optical Communication*, 2nd edn., Topics Appl. Phys., Vol.39 (Springer, Berlin, Heidelberg 1982) Chap.2

1.121 K.J. Ebeling: *Integrated Optoelectronics. Waveguides, Photonics, Semiconductors* (Springer, Berlin, Heidelberg 1993)

1.112 K. Fukagai, S. Ishikawa, K. Endo, H. Fujii, T. Yuasa: Jpn. J. Appl. Phys. 30, L371 (1991)

1.122 M.C. Wu, Y.K. Chen, M. Hong, J.P. Mannaerts, M.A. Chin, A.M. Sergent: Appl. Phys. Lett. 59, 1046 (1991)

1.124 M.A. Haase, J. Qui, J.M. Depuydt, H. Cheng: Appl. Phys. Lett. 59, 1272 (1991)

1.124 M. Ozawa, H. Okumima, Y. Morinaga, F. Hiei, K. Akimoto: 13th IEEE Semicond. Laser, Takamatsu (1992) Post-deadline paper, No. PD-16, 31

Chapter 2

2.1 F.H. Pollark, C.W. Higginbotham, M. Cardona: Int'l Conf. Phys. Semicond. Kyoto (1966) p. 20
P. Yu, M. Cardona: *Fundamentals of Semiconductors: Physics and Materials Properties* (Springer, Berlin, Heidelberg 1995)

2.2 D. Long: *Energy Bands in Semiconductors* (Wiley, New York 1968)

2.3 S.M. Sze: *Physics of Semiconductor Devices* (Wiley, New York 1969)

2.4 H.C. Casey Jr., M.B. Panish: *Heterostructure Lasers* (Academic, New York 1978) Pt.A, Chap. 4

2.5 M.A. Afromowitz: J. Appl. Phys. **44**, 1292 (1973)

2.6 K. Iga: *Laser Optics* (Ohm-sha, Tokyo 1988) (in Japanese)

2.7 H.K.V. Lotsch, F. Schröter: Das Laser-Farbfernsehen. Laser **2**, 37-39 (December 1970)

2.8 R.N. Hall, G.E. Fenner, J.D. Kingsley, T.J. Soltys, R.O. Carlson: Phys. Rev. Lett. **9**, 366 (1962)

2.9 T.M. Quist, R.H. Rediker, R.J. Keyes, W.E. Krag, B. Lax, A.L. McWhorter, H.J. Zeiger: Appl. Phys. Lett. **1**, 91 (1962)

2.10 M.I. Nathan, W.P. Dumke, G. Burns, F.H. Drill Jr., G. Lasher: Appl. Phys. Lett. **1**, 62 (1962)

2.11 I. Hayashi, M.B. Panish, P.W. Foy, S. Sumski: Appl. Phys. Lett. **17**, 109 (1970)

2.12 F.P. Kapron, D.B. Keck, R.D. Maurer: Appl. Phys. Lett. **17**, 423 (1970)

2.13 H.C. Casey Jr., D.D. Sell, M.B. Panish: Appl. Phys. Lett. **24**, 63 (1974)

2.14 R.E. Fern, A. Onton: J. Appl. Phys. **42**, 3499 (1971)

2.15 W.W. Chow, S.W. Koch, M. Sargent III: *Semiconductor-Laser Physics* (Springer, Berlin, Heidelberg 1994)

2.16 Y. Suematsu, K. Iga: *Introduction to Optical Fiber Communication*, 3rd edn. (Ohm-sha, Tokyo 1994)

2.17 K.C. Kao, G.A. Hockham: IEEE Proc. **113**, 1151 (1966)

2.18 M. Horiguchi, H. Osanai: Electron Lett. **12**, 310 (1976)

2.19 T. Miya, Y. Terunuma, T. Hosaka, T. Miyashita: Electron. Lett. **15**, 106 (1979)

2.20 H. Yokota, H. Kanamori, Y. Ishiguro, G. Tanaka, S. Tanaka, H. Tanaka, M. Watanabe, S. Suzuki, K. Yano, M. Hoshikawa, H. Shimba: OFC'86 (Atlanta) Paper PD-3

2.21 G.P. Agrawal, N.K. Dutta: *Long-Wavelength Semiconductor Lasers* (Van Nostrand Reinhold, New York 1986) p. 85

2.22 C. Flesia, P. Schwendimann (guest eds.): *Multiple-Scattering LIDAR Experiments* Appl. Phys. B **60**, No.4 (1995)

2.23 H.K. Choi, S.J. Eglash: CLEO'92 (Anaheim, CA) Paper CMF6, 36

Chapter 3

3.1 Zh.I. Alferov, V.M.Andreev, E.L. Portnoid, M.K. Trukan: Fiz. Tekh. Poluprov. **3**, 1328 (1969)

3.2 I. Hayashi, M.B. Panish, P.W. Foy, A. Sumski: Appl. Phys. Lett. **17**, 109 (1970)

3.3 R.L. Anderson: IBM J. Res. Dev. **4**, 283 (1960); Solid-State Electron. **5**, 341 (1962)

3.4 R. Dingle: *Festkörper-Probleme XV –Advances in Solid State Physics* (Vieweg, Braunschweig 1975) p. 21

3.5   H. Okumura, S. Misawa, S. Yoshida, S. Gonda: Appl. Phys. Lett. **46**, 377 (1985)
3.6   R.C. Miller, D.K. Kleinman, A.C. Gossard: Phys. Rev. B **24**, 7085 (1984)
3.7   S. Forrest, R.P. Schmidt, R.B. Wilson, M.L. Kaplan: Appl. Phys. Lett. **45**, 1199 (1984)
3.8   H.C. Casey Jr., M.B. Panish: *Heterostructure Lasers* (Academic, New York 1978) Pt.A, Chap.4
3.9   W.B. Joyce, R.W. Dixon: Appl. Phys. Lett. **31**, 354 (1977)
3.10   F. Ermanis, K. Wolfstirn: J. Appl. Phys. **37**, 1963 (1966)
3.11   H.C. Casey Jr., M.B. Panish: *Heterostructure Lasers* (Academic, New York 1978) Pt.A, p.221
3.12   H.K.V. Lotsch: Optik **32**, 116, 189, 299, 553 (1970/71)
3.13   H.C. Casey Jr., M.B. Panish: *Heterostructure Lasers* (Academic, New York 1978) Pt.A, p.51
3.14   Y. Itaya, Y. Suematsu, S. Katayama, K. Kishino, S. Arai: Jpn J. Appl. Phys. **18**, 1795 (1979)
3.15   T. Ikegami: IEEE J. QE-**8**, 470 (1972)
3.16   M. Asada, A. Adams, K. Stubkjaer, Y. Suematsu, Y. Itaya, S. Arai: IEEE J. QE-**17**, 611 (1981)

Chapter 4

4.1   C. Kittel: *Introduction to Solid State Physics* (Wiley, New York 1976)
       H. Ibach, H. Lüth: *Solid-State Physics*, 2nd edn. (Springer, Berlin, Heidelberg 1995)
4.2   Catalog of III-V Compound Semiconductors, Hitachi Cable Co., Ltd. (1985)
4.3   C.J. Neuse, G.H. Olsen: Appl. Phys. Lett. **26**, 528 (1975)
4.4   T. Mizutani, M. Yoshida, A. Usui, H. Watanabe, T. Yuasa, I. Hayashi: Jpn. J. Appl. Phys. **19**, L113 (1980)
4.5   W.H. Petzke, V. Gottschlach, E. Butter: Kristall und Technik **9**, 477 (1974)
4.6   M. Razheghi, J.C. Bouley, K. Kazmierski, M. Papuchon, B. de Cremoux, J.P. Duchemin: 9th IEEE Semiconductor Laser Conf., Rio de Janeiro (1984) Paper A-4
4.7   R.D. Dupuis, P.D. Dapkus: Appl. Phys. Lett. **31**, 839 (1977)
4.8   W.T. Tsang: IEEE J. QE-**20**, 1119 (1984)
4.9   B.I. Miller, J.H. McFee, R.J. Martin, P.K. Tien: Appl. Phys. Lett. **33**, 44 (1978)
4.10   W.T. Tsang: Appl. Phys. Lett. **45**, 1234 (1984)
4.11   M.B. Panish, H. Temkin: *Gas Source Molecular Beam Epitaxy*, Springer Ser. Mater. Sci., Vol.26 (Springer, Berlin, Heidelberg 1993)

Chapter 5

5.1   K. Iga: *Laser Optics* (Ohm-sha, Tokyo 1988) Chap.12 (in Japanese)
5.2   J.J. Hsieh: J. Cryst. Growth **27**, 49 (1974)
5.3   J.J. Hsieh: Proc. 6th Int'l Symp. on GaAs and Related Compounds, St. Louis, MI (1976) p.74
5.4   Y. Motegi: CTaInAsP/InP surface emitting lasers with short cavity length. Master Thesis, Tokyo Inst. of Technology (1982)

5.5　M.B. Panish, I. Hayashi: *Applied Solid State Science* 4, 235 (Academic, New York 1974)

5.6　H.C. Casey Jr., M.B. Panish: *Heterosctructure Lasers* (Academic, New York 1978) Pt.B, Chap.6

5.7　A.S. Jordan: J. Electrochem. Soc. 118, 781 (1971)

5.8　D.R. Ketchow: J. Electrochem. Soc. 121, 1237 (1974)

5.9　F.E. Rosztoczy, K.B. Wolfstirn: J. Appl. Phys. 44, 2659 (1973)

5.10　M.B. Panish: J. Appl. Phys. 44, 2659 (1973)

5.11　H.C. Casey Jr., M.B. Panish, K.B. Wolfstirn: J. Phys. Chem. Solids 32, 571 (1971)

5.12　C.S. Kang, P.E. Greene: *Proc. 1968 Symp. on Gallium Arsenide* (Inst. of Phys., London, 1969) p. 18

5.13　M.G. Milividskii, O.V. Pelevin: Inorg. Mater. 3, 1024 (1967) [Translated from: Izv. Akad. Nauk SSSR, Neorg. Mater. 3, 1159 (1967)]

5.14　H.C. Casey, Jr., M.B. Panish: J. Cryst. Growth 13/14, 818 (1972)

5.15　G.B.H. Thompson: *Physics of Semiconductor Laser Devices* (Wiley, Chichester 1980)

5.16　R.L. Moon, G.A. Antypas, L.W. James: J. Electron. Mater. 3, 635 (1974)

5.17　R.E. Nahory, M.A. Pollack, W.D. Johnstone Jr., R.L. Barns: Appl. Phys. Lett. 33, 659 (1978)

5.18　T. Yamamoto: 1.3 μm InGaAsP/InP semiconductor lasers. Dr. Eng. Thesis, Tokyo Institute of Technology (1981)

5.19　H.C. Casey Jr., M.B. Panish: *Heterostructure Lasers* (Academic, New York 1978) Pt.B p.39

5.20　S. Arai, Y. Suematsu: IEEE J. QE-16, 197 (1980)

5.21　M.A. Pollack, R.E. Nahory, J.C. DeWinter, A.A. Ballman: Appl. Phys. Lett. 33, 314 (1978)

5.22　H. Nagai, Y. Noguchi: Appl. Phys. Lett. 32, 234 (1978)

5.23　M.G. Astles, F.G.H. Smith, E.W. Williams: J. Electrochem. Soc. 120, 1750 (1973)

5.24　D.A. Pinnow, A.L. Gentile, A.G. Standlee, A.J. Timper, L.M. Hobrock: Appl. Phys. Lett. 33, 28 (1978)

5.25　L.G. Van Uitert, S.M. Wemple: Appl. Phys. Lett. 33, 57 (1978)

5.26　C.H.L. Goodman: IEEE J. Sol. State Electron. Dev. 2, 129 (1978)

5.27　S. Yamamoto, H. Hayashi, T. Hayakawa, N. Miyauchi, S. Yano, T. Hijikata: IEEE J. QE-19, 1009 (1983)

5.28　I. Ohta, M. Kazumura, I. Teramoto: LPE growth of $In_{1-x}Ga_xP/Ga_{1-y}Al_y As$ on GaAs. Proc. Int'l Symp. GaAs and Related Compounds, No.63, 54 (1981)

5.29　K. Kishino, Y. Koizumi, A. Yokochi, S. Kinoshita, T. Tako: Jpn. J. Appl. Phys. 23, L740 (1984)

5.30　S. Kaneiwa, H. Takiguchi, T. Hayakawa, S. Yamamoto, H. Hayashi, S. Yano, T. Hijikata: Appl. Phys. Lett. 46, 455 (1985)

5.31　K. Kishino, A. Harada, Y. Kaneko: IEEE J. QE-25, 180 (1987)

5.32　S. Mukai, H. Yajima, J. Shimada: Jpn. J. Appl. Phys. 29, L729 (1981)

5.33　H. Kawanishi, N. Tsuchiya: J. Appl. Phys. 58, 37 (1985)

5.34　T. Iwamoto, K. Mori, M. Mizuta, H. Kukimoto: Jpn. J. Appl. Phys. 24, L131 (1985)

5.35　K. Wakao, H. Nishi, S. Isozumi, S. Ohsaka, T. Kusunoki, I. Ushijima: Electron. Lett. 20, 374 (1984)

5.36　H. Kawanishi, T. Aota: Electron. Lett. 29, 263 (1984)

5.37　Y. Kawamura, H. Asahi, H. Nagai, T. Ikegami: Electron. Lett. 19, 163 (1983)

5.38 K. Kobayashi, S. Kawata, A. Gomyo, I. Hino, T. Suzuki: Electron. Lett. **21**, 931 (1985)

5.39 M. Ikeda, Y. Mori, H. Sato, K. Kaneko, N. Watanabe: Appl. Phys. Lett. **47**, 1027 (1985)

5.40 M. Ishikawa, Y. Ohba, H. Sugawara, M. Yamamoto, T. Nakanishi: Appl. Phys. Lett. **48**, 207 (1986)

5.41 N. Kobayashi, Y. Horikoshi, C. Uemura: Jpn. J. Appl. Phys. **19**, L30 (1980)

5.42 A.E. Bockharev, L.M. Dolginov, A.E. Drakin, L.V. Druzhinina, P.E. Eliseev, B.N. Sverdlov: Sov. J. Quantum Electron. **15**, 869 (1985)

5.43 C. Caneau, A.K. Srivastava, A.G. Dentai, J.L. Zyskind, M.A. Pollack: Electron. Lett. **21**, 815 (1985)

5.44 G.P. Agrawal, N.K. Dutta: *Long-Wavelength Semiconductor Lasers* (Van Nostrand Reinhold, New York 1986) Chap. 11

5.45 A.P. Shotov, O.I. Davarashvili: Izv. Akad. Nauk SSSR Neorg. Mater **13**, 610 (1977)

5.46 A.H. Groves, K.S. Nill, A.J. Strauss: Appl. Phys. Lett. **25**. 331 (1974)

5.47 L.R. Tomasetta, C.G. Fonstad: Apl. Phys. Lett. **24**, 567 (1974)

5.48 J.T. Longo, J. Harris, E. Gertner, J. Chu: J. Cryst. Growth **15**, 107 (1972)

5.49 J.L. Zyskind, C. Caneau, T.E. Glover, J.W. Sulhoff, C.A. Burrus, J.C. Centanni, A.G. Dentai, M.A. Pollack: Proc. 11th IEEE Int'l Semiconductor Laser Conf., Boston, MA (1988) Paper E-4

Chapter 6

6.1 H.M. Manasevit: Appl. Phys. Lett. **12**, 156 (1968)

6.2 H.M. Manasevit, W.I. Simpson: J. Electrochem. Soc. **120**, 135 (1973)

6.3 F. Koyama, H. Uenohara, T. Sakaguchi, K. Iga: Jpn. J. Appl. Phys. **26**, 1077 (1987)

6.4 P.D. Dapkus: J. Crystal Growth **68**, 345 (1984)

6.5 J.R. Arthur: J. Appl. Phys. **39**, 4032 (1968)

6.6 A.Y. Cho: J. Vac. Sci. Technol. **16**, 275 (1979)

6.7 A.Y. Cho: Thin Solid Films **100**, 291 (1983)

6.8 M.A. Herman, H. Sitter: *Molecular Beam Epitaxy*, 2nd edn., Springer Ser. Mater. Sci., Vol. 7 (Springer, Berlin, Heidelberg 1995)
The Technology and Physics of Molecular Beam Epitaxy, ed. by E.H.C. Parker (Plenum, New York 1985)

6.9 C.E.C. Wood: III-V growth by molecular beam epitaxy, in *GaInAsP Alloy Semiconductors*, ed. by T.P. Peasall (Wiley, New York 1982)

6.10 R. Dingle, A.C. Gossard, W. Wiegmann: Phys. Rev. Lett. **34**, 1327 (1975)

6.11 W.T. Tsang: IEEE J. QE-20, 1119 (1984)

6.12 J.H. Marsh, J.S. Roberts, P.A. Claxton: Appl. Phys. Lett. **46**, 1161 (1985)

6.13 M. Ilegems: Properties of III-V layers, in *The Technology and Physics of Molecular Beam Epitaxy* (Plenum, New York 1985)

6.14 M. Bafleur, A. Munoz-Yagne, A. Rocher: J. Crystal Growth **59**, 531 (1982)

6.15 T. Ito, M. Sinohara, Y. Imamura: Jpn. J. Appl. Phys. **23**, L524 (1984)

6.16 W.T. Tsang, F.K. Reinhart, J.A. Ditzenberger: Appl. Phys. Lett. **41**, 1094 (1982)

6.17 H.M. Manasevit: Appl. Phys. Lett. **12**, 156 (1968)

6.18 H.M. Manasevit, W.I. Simpson: J. Electrochem. Soc. **120**, 135 (1973)

6.19 H. Beneking, H. Rochle: J. Crystal Growth **55**, 79 (1981)

6.20 J.P. Hirtz, J.P. Larivain, D. Legnen, M. Razeghi, J.P. Puchemin: Low-pressure metalorganic growth and characterization of $Ga_x In_{1-x} As_y P_{1-y}$ on InP substrates, in *Gallium Arsenide and Related Compounds 1980* Conf. Ser. 56 (Institute of Physics, Bristol, UK 1981) pp. 29-35

6.21 C.R. Stanley, R.F.C. Farrow, P.W. Sullivan: MBE of InP and other P-containing compounds, in *The Technology and Physics of Molecular Beam Epitaxy*, ed. by E.H.C. Parker (Plenum, New York 1985)

6.22 T.H. Chiu, W.T. Tsang, E.F. Schubert, E. Agyekun: Appl. Phys. Lett. 51, 1109 (1987)

6.23 W.T. Tsang: J. Vac. Sci. Technol. B 3, 666 (1985)

6.24 W.T. Tsang: E.F. Schubert, T.H. Chiu, J.E. Cunningham, E.G. Barkhardt, J.A. Ditzenberger, E. Agyekun: Appl. Phys. Lett. 51, 761 (1987)

6.25 A.R. Calawa: Appl. Phys. Lett. 33, 1020 (1978)

6.26 K. Iga, F. Koyama, S. Kinoshita: IEEE J. QE-24, 1845 (1988)

6.27 F.J. Morris, H. Fukui: J. Vac. Sci. Technol. 11, 506 (1974)

6.28 M.B. Panish: J. Electrochem. Soc. 127, 2729 (1980)

6.29 A.R. Calawa: Appl. Phys. Lett. 38, 701 (1981)

6.30 M.B. Panish, S. Sumski: J. Appl. Phys. 55, 3571 (1984)

6.31 M.B. Panish, H. Tamkin, S. Sumski: J. Vac. Sci. Technol. B 3, 657 (1985)

6.32 M.B. Panish: J. Crystal Growth 81, 249 (1987)

6.33 M.B. Panish, H. Temkin: *Gas Source Molecular Beam Epitaxy*, Springer Ser. Mater. Sci., Vol. 26 (Springer, Berlin, Heidelberg 1993)

6.34 E. Venhoff, W. Pletschen, P. Balk, H. Lüth: J. Crystal Growth 55, 30 (1981)

6.35 T.H. Chiu, W.T. Tsang, J.A. Ditzenberger, C.W. Tu, F. Ren, C. Wu: J. Electron. Mater. 17, 217 (1988)

6.36 W.T. Tsang, E.F. Schubert: Appl. Phys. Lett. 49, 220 (1986)

6.37 W.T. Tsang: IEEE J. QE-23, 936 (1987)

6.38 W.T. Tsang: Appl. Phys. Lett. 50, 63 (1987)

6.39 J.C. Campbell, W.T. Tsang, G.J. Qua, B.C. Johnson: IEEE J. QE-24, 496 (1988)

6.40 E. Tokumitsu, Y. Kudou, M. Konagai, K. Takahashi: J. Appl. Phys. 55, 3163 (1984)

6.41 S. Yoshida, S. Misawa, A. Itoh: Appl. Phys. Lett. 26, 461 (1975)

6.42 T.H. Chiu, W.T. Tsang, J.E. Cunningham: J. Appl. Phys. 62, 2302 (1987)

6.43 Y. Kawaguchi, H. Asahi, H. Nagai: Oyo Buturi 56, 1178 (1987)

6.44 M. Konagai: Oyo Buturi 57, 1666 (1988)

6.45 G.J. Davies, D. Williams: In *The Technology and Physics of Molecular Beam Epitaxy*, ed. by E.H.C. Parker (Plenum, New York 1985)

6.46 A.Y. Cho, H.C. Casey, Jr., C. Radice, P.W. Foy: Electron Lett. 16, 72 (1980)

6.47 S. Nagata, T. Tanaka: J. Appl. Phys. 48, 940 (1977)

6.48 K.Y. Cheng, A.Y. Cho, W.R. Wagner, W.A. Bonner: J. Appl. Phys. 52, 1015 (1981)

6.49 W.T. Tsang: J. Crystal Growth 81, 261 (1987)

6.50 T.K. Uchida, T. Uchida, K. Mise, F. Koyama, K. Iga: Jpn. J. Appl. Phys. 29, L2146 (1990)

6.51 T.K. Uchida, T. Uchida, K. Mise, F. Koyama, K. Iga: 1st Int'l Meeting on Advanced Processing and Characterization Technologies. Tokyo (1989) Paper P-1

6.52 T.K. Uchida, T. Uchida, K. Mise, F. Koyama, K. Iga: Int'l Conf. on CBE and Related Growth Technique. Houston, TX (1989)

6.53 T.K. Uchida, T. Uchida, K. Mise, F. Koyama, K. Iga: Bull. P.M.E., No. 64, Tokyo Inst. of Techn. (1989) p. 11

Chapter 7

7.1 K. Kohra, S. Kikuta: *Technology of X-Ray Diffraction* (Publish. Serv., Tokyo University, Tokyo 1984) (in Japanese)
7.2 J. Hornstra, W.J. Bartels: J. Cryst. Growth **44**, 513 (1978)
7.3 T. Katohda: *Semiconductor Epitaxial Technology* (Sangyo-Tosho, Tokyo 1982) (in Japanese)
7.4 L.B. Valades: Proc. IRE **42**, 420 (1954)
7.5 T. Chakraborty, P. Pietiläinen: *The Quantum Hall Effects −Fractional and Integral*, 2nd edn., Springer Ser. Solid-State Sci., Vol. 85 (Springer, Berlin, Heidelberg 1995)
7.6 A. Kunioka, K. Kamimura: *Basic Semiconductor Engineering* (Asakura, Tokyo 1985) (in Japanese)
7.7 S.M. Sze: *Physics of Semiconductor Devices* (Wiley, New York 1969) p. 44

Chapter 8

8.1 T. Sanada, O. Wada: Jpn. J. Appl. Phys. **19**, L491 (1980)
8.2 N. Yokoyama, S. Ohkawa, H. Ishikawa: Jpn. J. Appl. Phys. **14**, 1071 (1975)
8.3 H.C. Casey Jr., M.B. Panish: *Heterostructure Lasers* (Academic, New York 1978) Pt.A, Chap. 4
8.4 Y. Suematsu: IEEE Proc. **71**, 692 (1984)
8.5 K. Peterman: 7th Europ. Conf. Optical Commun. No.10, 1 (1981)
8.6 K.Y. Lau, P.L. Derry, A. Yariv: Appl. Phys. Lett. **52**, 88 (1989)
8.7 K. Hamada, M. Wada, H. Shimizu, M. Kume, A. Yoshikawa, F. Tajiri, K. Itoh, G. Kano: 9th IEEE Semiconductor Laser Conf., Rio de Janeiro (1984) Paper C-1, p.34
8.8 M. Sakamoto, D.F. Welch, J.G. Endriz, D.R. Scifres, W. Streifer: Appl. Phys. Lett. **54**, 2299 (1989)
8.9 Y. Suematsu, M. Yamada: Proc. IECE Japan 57-C, 434 (1975)

Chapter 9

9.1 R. Ulrich, R.J. Martin: Appl. Opt. **9**, 2077 (1971)
9.2 K. Kishino, S. Kinoshita, S. Konno, T. Tako: Jpn. J. Appl. Phys. **22**, L473 (1983)
9.3 K. Aiki, M. Nakamura, T. Kuroda, J. Umeda: Appl. Phys. Lett. **48**, 649 (1977)
9.4 K. Iga, Y. Suematsu: 1st Europ. Conf. on Integr. Opt., London (1981) p. 70
9.5 H. Kogelnik, C. Shank: J. Appl. Phys. **43**, 2323 (1972)
9.6 K. Utaka, S. Akiba, K. Sakai, Y. Matsushima: IEEE J. QE-22, 1042 (1986)
9.7 K. Utaka, K. Kobayashi, F. Koyama, Y. Abe, Y. Suematsu: Electron. Lett. **17**, 983 (1981)
9.8 Y. Tohmori, H. Oohashi, T. Kato, S. Arai, K. Komori, Y. Suematsu: Electron. Lett. **22**, 138 (1986)
9.9 K. Komori, S. Arai, Y. Suematsu, I. Arima, M. Aoki: IEEE J. **25**, 1235 (1989)
9.10 S. Iida, K. Ito. J. Electrochem. Soc. **118**, 768 (1971)
9.11 Y. Tarui, Y. Komiya, Y. Harada: J. Electrochem. Soc. **118**, 118 (1971)
9.12 T. Kambayashi, C. Kitahara, K. Iga: Jpn. J. Appl. Phys. **19**, 79 (1980)

9.13 K. Kishino, Y. Suematsu, Y. Takahashi, T. Tanbun-Ek, Y. Itaya: IEEE J. QE-16, 160 (1980)
9.14 J.J. Hsieh, C.C. Shen: Appl. Phys. Lett. 30, 429 (1977)
9.15 Y. Itaya, T. Tanbun-Ek, K. Kishino, S. Arai, Y. Suematsu: Jpn. J. Appl. Phys. 19, L141 (1980)
9.16 S. Yamamoto, Y. Kurata, S. Matsui, T. Hayakawa, S. Yano, T. Hijikata: Papers of Tech. Group of IECE Electron. Devices ED 79-50, 49 (1979)
9.17 R.D. Burnham, D.R. Scifres: Appl. Phys. Lett. 27, 510 (1975)
9.18 T. Sugino, K. Itoh, M. Wada, H. Shimizu, I. Teramoto: IEEE J. QE-15, 714 (1979)
9.19 H. Ishikawa, H. Imai, T. Tanahashi, Y. Nishitani, M. Takusagawa, K. Tanahei: Electron. Lett. 17, 465 (1981)
9.20 A. Doi, N. Chinone, K. Aiki, R. Ito: Appl. Phys. Lett. 34, 393 (1979)
9.21 K. Moriki, K. Wakao, K. Kitamura, K. Iga, Y. Suematsu: Jpn. J. Appl. Phys. 15, 293 (1976)
9.22 I. Mito, M. Kitamura, Ken. Kobayashi, S. Murata, M. Seki, Y. Odagiri, H. Nishimoto, M. Yamaguchi, Ko. Kobayashi: J. Lightwave Tech. LT-1, 195 (1983)
9.23 Z. Liau, J. Walpole: Appl. Phys. Lett. 40, 568 (1982)
9.24 Y. Hirayama, H. Furuyama, H. Okuda: Int'l Symp. GaAs and Related Compounds, Karuizawa (1985)
9.25 B. Broberg, F. Koyama, Y. Tohmor, Y. Suematsu: Electron. Lett. 20, 692 (1984)
9.26 Z.L. Liau, J.N. Walpole: Appl. Phys. Lett. 46, 115 (1985)
9.27 Z.L. Liau, J.N. Walpole, V. Diadiuk: 11th IEEE Int'l Semiconductor Laser Conf., Boston, MA (1988) Paper N-4, p.168
9.28 S. Arai, M. Asada, Y. Suematsu, Y. Itaya, T. Tanbun-Ek, K. Kishino: Electron. Lett. 16, 349 (1980)
9.29 S. Kinoshita, K. Iga: IEEE J. QE-23, 882 (1987)
9.30 A. Ibaraki, K. Kawashima, K. Furusawa, T. Ishikawa, T. Yamaguchi, T. Niina: Integ. Opt. and Opt. Fib. Commun., Kobe (1989) Paper 18B1-3
9.31 P. Besomi, R. Wilson, W. Wagner, R. Nelson: J. Appl. Phys. 54, 535 (1983)
9.32 S. Takahashi, H. Nagai: J. Cryst. Growth 51, 502 (1981)
9.33 J. Kinoshita, H. Okuda, Y. Uematsu: Electron. Lett. 19, 215 (1983)
9.34 M. Asada, A. Kameyama, Y. Suematsu: IEEE J. QE-20, 745 (1984)
9.35 Y. Sasai, N. Hase, M. Ogura, T. Kajiwara: J. Appl. Phys. 59, 28 (1986)
9.36 N. Dutta, D. Craft, S. Napholtz: Appl. Phys. Lett. 46, 123 (1985)
9.37 A. Chailertvanitkul, K. Iga, K. Moriki: Electron. Lett. 21, 303 (1985)

Chapter 10

10.1 H. Soda, K. Iga, C. Kitahara, Y. Suematsu: Jpn. J. Appl. Phys. 18, 2329 (1979)
10.2 K. Iga, F. Koyama, S. Kinoshita: IEEE J. QE-24, 1845 (1988)
10.3 I. Watanabe, F. Koyama, K. Iga: Jpn. J. Appl. Phys. 16, 1598 (1988)
10.4 M. Oshikiri, H. Kawasaki, F. Koyama, K. Iga: Photon. Tech. Lett. 1, 11 (1989)
10.5 A. Kasukawa, Y. Imajo, T. Fukushima, H. Okamoto: 48th Device Res. Conf., Santa Barabara, CA (1990) Post Deadline Paper VB-2
10.6 K. Iga, S. Kinoshita, F. Koyama: Electron. Lett. 23, 134 (1986)
10.7 F. Koyama, H. Uenohara, T. Sakaguchi, K. Iga: Jpn. J. Appl. Phys. 26, 1077 (1987)

10.8 F. Koyama, K. Tomomatsu, K. Iga: Appl. Phys. Lett. **52**, 528 (1988)

10.9 F. Koyama, S. Kinoshita, K. Iga: Trans. IEICE of Jpn. E **71**, 1089 (1988)

10.10 F. Koyama, S. Kinoshita, K. Iga: Appl. Phys. Lett. **55**, 221 (1989)

10.11 M. Tanobe, F. Koyama, K. Iga: Electron. Lett. **25**, 1444 (1989)

10.12 T. Sakaguchi, F. Koyama, K. Iga: Electron. Lett. **24**, 928 (1988)

10.13 S. Uchiyama, K. Iga, Y. Kokubun: 12th Europ. Conf. on Optical Commun. (1986) p. 37

10.14 K. Iga, M. Oikawa, S. Misawa, J. Banno, Y. Kokubun: Appl. Opt. **21**, 3456 (1982)

10.15 S. Uchiyama, K. Iga: Electron. Lett. **21**, 162 (1985)

10.16 E. Ho, F. Koyama, K. Iga: MOC/GRIN'89, J2, 242 (1989)

10.17 A. Ibaraki, K. Kawashima, K. Furusawa, T. Ishikawa, T. Yamaguchi, T. Niina: Jpn. J. Appl. Phys. **28**, L667 (1989)

10.18 M. Shimada, T. Asaka, Y. Yamasaki, H. Iwano, M. Ogura, S. Mukai: Appl. Phys. Lett. **57**, 1289 (1990)

10.19 Y.H. Lee, B. Tell, K. Brown-Goebeleer, J.L. Jewell: 48th Dev. Res. Conf., Santa Barbara, CA (1990) Post Deadline Paper VB-1

10.20 J.L. Jewell, A. Scherer, S.L. McCall, Y.H. Lee, S.J. Walker, J.P. Harbison, L.T. Florez: Electron. Lett. **25**, 1123 (1989)

10.21 R.S. Geels, L.A. Coldren: 48th Dev. Res. Conf., Santa Barbara, CA (1990) Paper VIIIA-1

10.22 K. Iga, F. Koyama: *Surface Emitting Lasers* (Ohm-sha, Tokyo 1990) (in Japanese)

10.23 N.W. Carlson: *Monolithic Diode-Laser Arrays*, Springer Ser. Electron. Photon., Vol. 33 (Springer, Berlin, Heidelberg 1994)

Subject Index

Springer Series in *Materials Science*

Advisors: M.S. Dresselhaus · H. Kamimura · K.A. Müller
Editors: U. Gonser · R.M. Osgood, Jr. · M.B. Panish · H. Sakaki
Managing Editor: H. K.V. Lotsch

Springer-Verlag
and the Environment

We at Springer-Verlag firmly believe that an international science publisher has a special obligation to the environment, and our corporate policies consistently reflect this conviction.

We also expect our business partners – paper mills, printers, packaging manufacturers, etc. – to commit themselves to using environmentally friendly materials and production processes.

The paper in this book is made from low- or no-chlorine pulp and is acid free, in conformance with international standards for paper permanency.